MAN AT NATURE'S PINNACLE

A novel about the importance and
significance of man's interrelationships
with animals and plants,
as demonstrated in ancient Europe.

By

Moses Ekebuisi

Grosvenor House
Publishing Limited

All rights reserved
Copyright © Moses Ekebuisi, 2025

The right of Moses Ekebuisi to be identified as the author of this
work has been asserted in accordance with Section 78
of the Copyright, Designs and Patents Act 1988

The book cover is copyright to Moses Ekebuisi

This book is published by
Grosvenor House Publishing Ltd
Link House
140 The Broadway, Tolworth, Surrey, KT6 7HT.
www.grosvenorhousepublishing.co.uk

This book is sold subject to the conditions that it shall not, by way of
trade or otherwise, be lent, resold, hired out or otherwise circulated
without the author's or publisher's prior consent in any form of
binding or cover other than that in which it is published and
without a similar condition including this condition being
imposed on the subsequent purchaser.

This book is a work of fiction. Any resemblance to
people or events, past or present, is purely coincidental.

A CIP record for this book
is available from the British Library

ISBN 978-1-80381-630-2
eBook ISBN 978-1-80381-631-9

DEDICATION

This novel is dedicated to all those who play roles in facilitating nature conservation, human freedom and welfare, and peace of the world. Their efforts may seem unnoticed, unappreciated, and unrewarded; however, the message from this novel has illuminated the significance of their endeavours and revealed the subsequent hidden recompense. Their work is now visible as contributing to the protection, conservation, and care of the building blocks that make up and sustain the sacred universe and its order – a divine duty that earns rich and everlasting spiritual and eternal reward.

ACKNOWLEDGEMENT

I acknowledge Jim Newmark for the proof reading.

THE AUTHOR'S BIOGRAPHY

Moses Ekebuisi is a Biomedical Sciences graduate of the University of Bradford, in the United Kingdom. Born in Nigeria, he now lives in the UK.

Moses carried out a research with other scientists in the Clinical Oncology Unit at the University of Bradford. He is a co-author of the publication in the journal, *Annals of Oncology*.[1]

Moses also carried out a research with other scientists at the University of Leeds, investigating the effects of agricultural chemicals on some soil organisms. He is a co-author of the publication in the *Journal of Applied Entomology*.[2]

It was a combination of Moses' scientific background and mystical experience in agricultural science research

---

[1] Comparative in vitro chemosensitivity of DLD-1 human colon adenocarcinoma monolayers and spheroids to the+ novel indoloquinone EO9." R.M. Phillips, M.C. Bibby, B.P. Cronin, M. Ekebuisi - Ann Oncol, 1992.

[2] Effect of seed bed preparation, soil structure and release time on the toxicity of a range of grassland pesticides to the carabid beetle Pterostichus melanarius (Ill.)(Col., Carabidae) using a microplot technique." J.S. Bale, M. Ekebuisi, and C. Wright. *Journal of Applied Entomology* 113.1–5 (1992): 175–182.

that inspired him to write his first book, Interconnectedness of Life. It was the same experience that led to his writing this novel.

Moses is also a musician. He performs world music for audiences. Based on his research on the therapeutic benefits of musical rhythms, he now runs African drumming classes and relaxation sessions for improving well-being. This is in addition to his music performance. More information about his music and books can be accessed on his website, https://www.mosekemusic.co.uk

Email: moses@mosekemusic.co.uk

His books are available in all major book retailers.

INTRODUCTION

The novel is set in the prehistoric age in Europe. It draws on my scientific background and mystical experience, and tells the prophetic tale of a prosperous farmer who started using chemicals on the land – to his eventual cost. It was in the middle of his suffering that he received a mystical call that directed him what to do to recover from his agricultural disaster. By following the direction, he solved his problem and revealed to mankind the mysteries of nature, the position and role of man in nature, and the validity of the concept held by the mystics of the past. The concept is that no life exists in isolation, independent of the others.

I was inspired to write the novel after becoming aware that a concept held by the mystics of the past is in agreement with today's scientific findings. The concept is that all life is interconnected, that nothing exists in isolation and so, the real nature of individuals and events can only be correctly understood in the context of their connections with all others. This belief is held by Buddhists.

The belief is also found in the Gaia Theory. Gaia was the Greek goddess of the Earth. Her theory indicates that no life exists independently, but rather that all

Living things are interconnected, forming a system where through their interactions with each other and their surroundings, they ensure their well-being and survival.

This belief is also expressed in the Catholic Bible, Sirach 42: 24–25. To quote the expression: "All things exist in pairs, one opposite the other, and he made nothing that was incomplete. Each thing strengthens the good parts of the other; who can get enough of seeing God's glory?" This is an expression of mutualistic relationships between different organisms where through their interaction, each perfects the well-being, life and survival of the other.

So, the belief held by different mystics of the past totally agrees that all life evolved together to interact with each other and through this, they ensure their survival. Today, this belief has been proved valid by modern science. I experienced this in one of my scientific research projects in agricultural science. (The details and data of the research are in my first published book, *Interconnectedness of Life*). The research took place at the Leeds University farm in the United Kingdom.

The aim of the research was to investigate the effect of agricultural pesticides on the soil organisms. We observed that on many occasions, in the plots where the pesticides were applied, the grass yield was higher than in the plots where they were not applied. We thought that the grass high yield in the plots with the pesticides was because the pesticides killed pests in the grass, leading to the better yield. But we were wrong.

The population of the pests in the plots with the pesticides was not significantly less than the population in the plots without the pesticides. So, this occurrence became a phenomenon beyond our explanation. It was in the middle of this research that I had a mystical experience.

One night, I was in a deep sleep. It was like I was in a trance state. Then I heard the head of the project calling me in a loud, pressing and excited voice. On my answering the call, he raised up a sheet of white paper which was shining like a dazzling sun. He told me that it was on the paper that the reason behind the high grass yield in the plots with the pesticides was written. From where I was, I could not see anything written on the paper except the dazzling white light that was coming from it. So, I attempted to move closer to him, but he told me in a loud and powerful voice that I should not come close or attempt to touch the paper because it was too powerful for me. He said that instead I should stay where I was and listen to a voice which was to read out what was written on the paper. So, I listened and suddenly saw the Leeds University farm secretary. In a gentle, clear voice, she read out this:

"The animals supply the needs of man;
the plants supply the needs of the animals;
for all the soil organisms, bits and pieces assembled together,
mimic the ecosystem in order to supply the needs of the plants."

When I woke up, I pondered on the ecosystem mentioned in the information. The ecosystem is a system of interdependent organisms, a community of organisms where the organisms interact in an interdependent manner and through this, they ensure their well-being and survival, as well as ensuring the required conditions for the continuity of the whole system.

In every field crop, there is a group of organisms known as the natural enemies. They kill the other organisms including the crop pests, or they reduce the reproductive potential or the number of the other organisms. Through these ways, the natural enemies regulate the populations of the other organisms, maintaining the populations at the required levels in the environment. The natural enemies include predators and parasites.

In the research, I was the one who was sampling the soil organisms and recording the numbers. So, I went to the laboratory where the data was stored. I looked closely at the number of different groups of organisms in the data. Usually, the population of many of the groups of the organisms in the plots treated with pesticide was reduced to be lower than the population of those in the plot without the pesticide. We were aware of this. But what we didn't notice was that the population of one particular group of the organisms known as oribatid mites was increasing as the population of the other groups was decreasing, and the more the population of the other groups decreased, the more the population of the oribatid mites increased. I also noticed that the more the population of the oribatid mites increased, the higher the grass yield. This means that the increase in the grass yield was connected to the increase in the population of the oribatid mites.

The mites play an important role in maintaining soil fertility. They depend on the plants for their food and in return the plants depend on the mites for nutrients that enable their growth and good yield. The food which the mites feed on is plant litter. The plants shed the litter, giving it to the mites. By feeding on the litter, the mites play an important role in decomposing the litter, releasing nutrients in them and putting the nutrients back into the soil for the plants to absorb through their roots. Today, scientists are using the mites in organic farm research. They found that the nutrients which the mites release from the litter are better than the conventional fertilizer which farmers use. For example, the nutrients enabled an okra plant to flower and bear fruit (pods) earlier and the pods were bigger than the ones obtained from an okra plant grown in a soil with the conventional fertilizer.

It also emerged that the increase in population of the mites in the plots with the pesticide was connected to the changes which the pesticide made in the population of the natural enemies. Just as the natural enemies control the population of the pests, ensuring that the population doesn't grow above the required level in the environment, they also control the population of the mites, maintaining it at a specific level. When the pesticide was applied, it remarkably reduced the number of the natural enemies, thereby giving the mites' population the freedom to rise above usual. Also, since the mites live under the litter, the litter protected them from the pesticide. These factors led to the increase in the grass yield. So, the increase in the grass yield was the result of chain of changes which the pesticide initiated – all organisms including the crop are interconnected and a change in one, leads to a change in the other.

This finding and other independent research validate the concept held by the mystics of the past: that no life exists in isolation; that all life is interconnected and that the real nature of individuals or events can only be correctly understood in the context of their connections with all others.

In his teaching in the Bible, 1 Corinthians, 12: 4-26; St Paul, one of the most important figures in the apostolic age revealed that modeling a religious community on the concept, leads to achievement of spiritual success: just as the organisms in the natural system have their unique characteristics which enable them to interact for perfecting and sustaining each other, that Christian community is a unitary body and the individuals in it have their own unique characteristics which are a variety of spiritual gifts. The spiritual gifts include a gift of wisdom, a gift of knowledge, a gift of prophecy, a gift of healing. He said that the particular way in which the gifts are distributed is for a good reason. He likened the individuals with the different gifts to different parts that form one unitary body where they function in an interdependent manner – an interrelationship where a function in which one part lacks the gift to carry it out, the other has; the success of one part becomes the success of the other part; if one part is hurt, all the parts are hurt with it. If one part is given special honour, all parts enjoy it. He stressed that the force that stimulates the function of the gifts is love and that one without love will not be willing to share his own gift, thereby making his gift useless.

St Paul also revealed that no matter how weak a gift of an individual may seem, the gift is indispensable to the existence of the whole system. A similarity of this can be seen in the forest which itself is also a system of interdependent organisms where seemingly weak plants like saprophytes and lianas play significant roles in the survival of the forest.

It is now obvious that the concept held by the mystics of the past, complements science, and the mystics perceived the concept through their religions. This then raised the question: Could it be that religion and science complement each other? A great scientist, Albert Einstein gave the answer by saying yes in his essay (Einstein's Essay "*Science and Religion*", published in 1954). To quote him: "Science without religion is lame, religion without science is blind."

The answer points out that religion, unless it is abused, may not be needless. It is an integral part of humanity which evolved for a good purpose. It is part of cultural practice across the whole world, where in one part of the world, one form of it exists, and in the other part, a similar form if not the same type exists. My consciousness of this fact led to my looking closer at the ancient religion in the culture where I originated. Although I am a British citizen, I came from the Igbo culture in the southeast of Nigeria. I noticed that the Greek goddess of Earth, Gaia was the same goddess which the ancient Igbos worshipped. In-fact both have the same name. The Greek word, Gaia means the land or earth. The name of the ancient Igbo goddess is Ala or Ani, depending on the dialect. She is the goddess of fertility and morality; the queen of all deities and the author of the law of the earth. Her name, Ala or Ani means the land or earth. (Ref: 'Things Fall Apart', by Chinua Achebe: *"Before it was dusk, Ezeani who was the priest of the earth goddess, Ani, called on Okonkwo in his Obi, 'The evil you have done can ruin the whole clan. The earth goddess whom you have insulted may refuse to give us her increase, and we will perish.... 'You will bring to the shrine of Ani tomorrow one she-goat, one hen, a length of cloth and a hundred cowries.'"*

The ancient Igbos also had stone cult similar to those found in ancient Europe. Expression of the cult is found in one of the sayings of the ancient Igbos: *'Agbakide nkume ogho alunsi'*, meaning, if stones are put together, they turn into a deity. If the ancient Igbos had an access to the Stonehenge in Europe, they would only see and interpret it as a giant stone cult for invincible deities.

With the above experience, I was inspired to write this novel. The setting is in the prehistoric age in Europe, when people were practising the ancient religion, when their laboratory was their religious monuments and when their mystical insight was very sharp. It was the period in which they lived very close to nature and their folk stories were about organisms in their natural environment. It was the period they used draft animals for ploughing their agricultural land and for transporting heavy loads.

The novel reveals the mysteries of nature, man's position in nature and his required role. It brings to light the meaning and validity of some of man's behaviour which might have been previously perceived as irrational. It unveils how science and religion complement each other.

Chapter 1

In the ancient world, at the heart of European land, stood the Floradale, a massive fertile landscape. It was a land where plants flourished perfectly, bearing their fruits in their season, and its animals were healthy, living on abundant nourishments. The habits of all living things there, and their interrelationships were in accordance with the creator's order.

In the land was a farmer, Ward. He was so successful in his occupation that his agricultural produce improved the lives of the neighbouring clans, hamlets and villages. Throughout the whole territory, he was highly revered for his intimate knowledge of forest medicine and the power of healing. The story about his power of healing, mystical qualities, and insight into the mysteries of nature travelled far and wide.

Although Ward lived in Floradale, it was not his original land. His native land was Highland, a community which was very far away from Floradale and which consisted of poor farmers. The soil of the land was very infertile because of overcultivation and overgrazing. The people hardly produced enough food to last beyond their immediate needs. It was this hard life that sparked off an incident that led to Ward's glorious status.

The one who suffered the hard life most was Kenelm, Ward's father. He was a weak farmer. After planting his crop, he always obtained poor harvest. In one harvest season, his agricultural ill-luck was so severe that it was perceived by the Highland community as a manifestation of a bad relationship between him and the gods of the community. After that season, Kenelm's barn was almost empty.

"Gladys," Kenelm called his wife sadly, "as you have seen, the harvest was the worst we have ever come across. I am sure the grain in the barn will not even last for long. There is virtually no winter feed for the sheep. Very soon they will die of starvation. I am going to the market to exchange two of the sheep for grain and add it to the meagre supply in the barn. If we use it carefully, it might last up to the beginning of the spring, when the life will once again return to the earth. Let's pray that in the spring, the sky god will send warm weather earlier so that I will be able to resume soon my search for anything edible in the forest. When I return from the market, I will start collecting seaweed for the rest of the sheep to live on. If there is any way you feel you can support me, please do."

Poor sickly lady, what could she do? All that she had to do was to dream, wish and pray. "May our gods heal and strengthen me to help," she implored. "May the spirits of our ancestors direct you and make your plan a success. May they keep Ward healthy, nurture him to grow up and join you in your effort. I entreat the earth goddess to show us her mercy by blessing our crop to yield abundantly in the next season."

"May our deities hear your prayer," Kenelm replied in invigorated mood. He then took the two sheep and went to the market.

Soon after Kenelm had returned from the market, he had been collecting seaweed while the people in the community were celebrating a festival. It was a religious festival as Highland had a cult of fertility deity. During the planting season, the people offered sercrifices to the deity as an appeasement to her to make their soil fertile. The festival was a period of thanksgiving. It took place after harvest. Gareth and Manley were very happy. Their happiness seemed to suggest that they had a good harvest. It seemed their offering to the goddess during the planting season was fully acknowledged. One could read the excitement in their mood as they strolled around the community, meeting their colleagues, conversing and joking with them. As they strolled, something suddenly caught their attention. It was a vague unusual image afar off.

"What is this?" Gareth asked in surprise.

"I don't know," replied Manley. They stood and looked at the object in a brief silencc. Then, they walked towards it. Closer and closer the image continued to be clearer until they saw what it was – Kenelm was returning from the seashore, carrying seaweed on his back. He was exhausted. He stooped as he walked, panting like a weary, thirsty dog.

The men had approached him. Blocking his way, Gareth mockingly said to him: "We collect seaweed for manuring our soil before the planting season, and not in this period. What are you carrying seaweed for at this time? When are you ever going to behave like a member of our community?" There was an outburst of derisive laughter from him and Manley.

"The fortunate one," Kenelm replied sorrowfully, his voice shaking with emotion, "you are right. Indeed, this period is the time for one to rest and enjoy his agricultural produce. It is not a time for him to suffer. Before the last planting season, I manured my soil, just as you did yours. During the season, I tilled the soil with my mattock and hoe and then sowed my crop. When the harvest season came, I took my sickle to reap what I sowed but only to find out that there was nothing there to reap. You, the fortunate one whom our gods listen to, please ask the gods why the common saying 'One reaps what he sows' does not apply to me. Go and tell them my misfortune and ask them why they allow it. Ask them to tell you if I offended them in any way, and if so, what should I do to placate them; to appease them to nullify this ill-luck, a punishment as dreadful as fatal infectious disease."

Like a dog's reaction to a sudden snap by a strange object which the dog is sniffing, in shock Gareth and Manley backed away from him, giving way. Kenelm staggered through as the men stood in horror, gaping at him.

"Are you surprised about what you saw and heard?" Manley whispered to Gareth. "How do you expect him to enjoy the festival? How can he not suffer when he offers nothing to the goddess? If you go to his home, you will hardly see any evidence of good offering to the deity. All that one can see there are just traces of grain paste."

"You are just talking about a sacrifice which he didn't offer," replied Gareth. "You haven't mentioned something worse than that which he did. You haven't mentioned why his wife is not able to help him, why he suffers alone."

"Is it as a result of our gods' anger against him?" Manley asked with a shudder.

"That is the story," Gareth continued. "He deliberately provoked the gods and brought infirmity upon his wife. This happened some years ago, when he was preparing to marry the wife. In his father's place were some antlers which the father dedicated to the god of the forest as his expression of thanks to the god for hunting success which the god made possible. But one day, after the father had died, Kenelm was to be visited by the newly betrothed girl. But he didn't have enough food for him and the girl. So, he took the sacred antlers to Harewood community and exchanged them for foodstuff from farmers who needed them for making their sickles. He returned, thinking that he only exchanged the antlers for foodstuff, but not knowing that he also exchanged them for his wife's ill health. It is the seed which he sowed that he is reaping but he doesn't know that."

"This is a new story to me," Manley replied in deep shock.

"Yes," said Gareth, "you didn't hear the story because those who know about it maintained their silence. Although everyone tried not to talk about it, what is happening these days is now forcing us to speak out."

The men spread the news in the whole community. Alger, the heartless priest heard it and affirmed that Kenelm's suffering was retribution from their gods.

Although Kenelm was aware that the community knew him as one who failed to appease their gods well, he was not willing to change his ways. He had his personal reason. He saw humanity as the most important thing to their gods, and his family was part of the humanity. He believed that everything their gods provided was for maintaining that humanity. He saw his wife and Ward, the only son, as a special gift from the gods, and his maintenance of them as the most sacred duty required of him. He was not willing to hand back to the gods the food he received from them in the name of offering, leaving his family to starve and die. To him, doing that was sacrilegious. The gods might need food from him, but that should be when the food would be plenty, when there would be a leftover to hand back to the gods. With this notion, he was confident that he was doing the right thing, and so paid no attention to what people thought about him.

In spite of all the difficulties, Kenelm managed to maintain his family and keep them happy. He was a good storyteller, and through this, he entertained them well. He could easily create interesting stories, and was able to modify uninteresting ones, making them very interesting. It was this skill that he applied after the poor harvest. He told a story that remained fresh in Ward's mind throughout his life.

It was winter night and the people of Highland were enjoying storytelling in their thatched huts. The darkness of the night swallowed every object except the huts whose existence was being announced by a reddish yellow light from log fires inside them. Again, and again people's voices rose in excitement and faded.

"They are excited over a badly told story," Kenelm said to his family, alluding to some of the community members in their poor storytelling. "I always laugh whenever I remember how Gareth, the affluent struggled to tell the story of nightingale, to the extent that his children helped him to complete it."

"Yes," replied the wife supportively, "that is one of the failings of those who are blessed with a good harvest. They got the blessing but lack the ability to tell a good story."

Kenelm was now about to tell the story, but first he revived their fire with more firewood, ensuring that the room was very warm. A story told in such atmosphere and that time of the season was sweeter than when told at any other period.

"Nightingale! Nightingale!" his voice rose in praise, "a symbol of beauty and melody. Your sweet song and impressive feathers have made you an icon. Everyone wishes to see you, hear your song for invigoration."

He paused. Ward was attentive, waiting expectantly to hear the story. Kenelm then took deep breaths before continuing.

"Some time ago," he went on, "there was a community of birds. The birds there lived in peace, loving each other. The thing they valued most was a song. Anyone that sang a good song, they respected and gave him gifts. So, one day, Nightingale left the community and travelled to a land very far away. There he listened to and imitated different types of sounds, ranging from the sounds of water, wind, rain; voices of different animals and shrill of insects. He then put them together and formed a complex sweet song. He returned to the community and started singing the song in an open space. When the other birds heard him, they flocked to the place listening to him. As they listened, they were amazed because the song was sweeter than their own songs. This provoked their envy. When Nightingale finished singing, he expected the birds to thank him and give him a present. Sadly, the birds laughed at him mockingly, booing, jeering. Disappointed and angry, he vowed not to sing for the birds anymore. Since then, he chooses to sing in a hidden place where he enjoys his song alone."

"That is why you hear a nightingale's song coming from a deep cover," Ward's mother added, turning to him, smiling fondly.

Ward was intrigued. "Father, when did this happen?" he asked.

"It happened in the dim past," Kenelm replied, "when tranquility and peace were the order of the day; a time when trees worked as gods' messengers, sheltering the birds and supporting weak plants, ensuring their survival; a time when the sound of the breeze was a soothing musical rhythm, which plants dance to, swaying, whistling, entertaining themselves."

The story and the answer were like a relaxing music to Ward. He had a good sleep that night. The following day, he met his playmates and told them the story. The children were refreshed and inspired to engage in new leisure activities. They developed a simple song about the bird. On many occasions, during the spring and summer period, they went to the forest, singing, asking if the nightingale had returned from his long journey, hiding somewhere in the forest, and if so, let him sing his song for them because they had some presents for him.

In winter, the children played in the snow. During the time, snow fell like a myriad of mini down feathers, forming heaps of brilliant soft snow on the earth. In excitement, the children ran out of their huts, whooping; scooping, compressing the snow and playfully throwing it at each other. They also molded the snow into crude and shapeless images of a nightingale. Engaged in the exciting activity, they forgot the cozy huts, only to return to them when the snow melted away.

It was in such an atmosphere that Ward was growing, enjoying the fun and fellowship. But soon, the enjoyment was going to give way to a sad belief – a belief that he and his family were ill-fated in the community. It was the torment of this belief that triggered a mystical occurrence that led to Ward's glorious status.

Chapter 2

Ward was growing up very fast, showing a keen interest in agriculture. By now he had been helping his father in crop planting and shepherding. One day, his father called him:

"Go to the sheep pen," he said. "Take the sheep to graze in the field. Stay there until I come to see you."

Ward ran to the sheep pen and led their five sheep to their barren field. As he was there, his playmate, Everard visited.

"Where is Ward?" the boy asked Kenelm who was repairing his henhouse.

"This is not a time for play," Kenelm growled at the boy. "Go to the field and help him to look after the sheep."

"Where is the field?" the boy asked in excitement. Kenelm showed him by pointing to the location of the field. The boy was enthusiastic to take part in the shepherding. He ran quickly to the field. Seeing the condition of the field and the sheep, his excitement was replaced by a shock.

The field was so barren that the grass in it was very sparse. The sheep were stunted and very bony.

This reminded the boy of a story which his father told him some time ago.

"Ward," the boy called, "this field and sheep are like the type in a story which my father told me."

"What is the story?" Ward asked with interest.

"Once upon a time," the boy began, "in a certain land, after a harvest season, people offered good sacrifices to the fertility goddess, thanking her for blessing their soil. But there was one man there who always offered poor sacrifice to the goddess. Because of that, the goddess cursed his soil and his field became as barren as… as… ah, ah," he stammered over his words as he failed to imitate his father's description of the barren field. Pointing at the field, he went on, "as barren as this, and his sheep as bony as these," he pointed at the sheep. "Eventually, the man and his wife died and their children inherited the curse. In the end, all the children died of starvation."

After the story, Ward's mood changed. Fear and anxiety had crept into his mind. He stood like a withered tree. He felt pity for the unfortunate family in the story and thought about his father's poor sacrifice to their deities. He was deeply apprehensive. His friend had read the extent of his despondency and so spoke to him in an attempt to enliven him. But it seemed that Ward was not hearing him. The friend could not bear the gloomy mood anymore and so he left him.

When Kenelm walked in, he noticed that Ward was weak and depressed. "What is the matter?" he asked.

"Is such simple job very difficult for you, to the extent that you look so exhausted?" He laughed, thinking that Ward was only showing signs of weakness that were always displayed by young people after long hours of husbandry work. "Don't worry," he comforted. "It is not unusual. It takes time to become used to the occupation." He took Ward by hand and led him and the sheep to their homestead.

Although Ward ate enough food and had full rest, his gloomy state remained. All the time, he thought about the sad story. He visualised the goddess in the story as the same in Highland. His father's poor offerings to the goddess always came to his mind. He feared that his family was destined to suffer like the family in the story. He felt that the only way to avoid the suffering was to engage in any other occupation other than agriculture. With this feeling, Ward's interest in agriculture continued to diminish. Every time agricultural duty was assigned to him, he disappointed the parents. The father was now fed up with his behaviour.

"I don't know what is wrong with Ward," Kenelm complained bitterly to his wife one night. "Where does this laziness come from? Yesterday, you saw how in his absent-mindedness, he left the sheep to wander away. He is weak in tilling the soil and lazy in collecting seaweed. I wonder what he is going to be. How is he going to feed himself when we depart this life?"

"I am also worried," replied the wife. "Let's pray that with time he will change and become a good farmer or he will develop skills in a non-agricultural occupation."

"Non-agricultural occupation, such as what?" Kenelm retorted.

"Hunting," replied the wife.

"Yes, hunting occupation, a good occupation like the occupation of Marshland people," Kenelm replied sarcastically, alluding to people of Marshland whom they saw as unfortunate hunters. "When I think about all these things that are happening to me, I wonder what I did against our gods."

Ward had gone to bed when the parents were discussing, but he was not yet asleep. On hearing the last statement, he jerked his head like a watchdog that sensed an intruder. He listened anxiously.

"The fertility goddess left my soil unblessed, leaving it barren," Kenelm continued. "But I thought that the interest Ward was showing in agriculture earlier was a good omen; that it was a sign that our gods have forgiven me for whatever they felt I did against them. I presumed that they blessed and gave us Ward as one who would put smiles on our faces; as one to be a good farmer, whom our gods will make our soil fertile for him. You can now see that his recent behaviour has proved me wrong. It is now obvious that he is going to be a weak farmer who is destined to inherit infertile soil."

On hearing this, Ward interrupted sharply: "Father! Why is our soil infertile?" He listened uneasily.

"I don't know!" Kenelm replied harshly. "Don't ask me. Ask the fertility goddess. Ask her what I have done for this punishment."

The message was clear and terrifying. The vague idea in Wards' mind was now a distinct truth. He saw his father's poor harvest as a result of a curse which the goddess placed on him. He had no doubt in his mind that he had inherited the curse. With this belief, Ward's lack of interest in agriculture continued to grow. The worry about his future was like an illness that tormented the parents.

It was likely that Ward's fear would fade away with time, when he grew up and matured enough to know that the story was a mere fiction and didn't actually happen. But the result of the ways his father dealt with things seemed to reinforce his fear, making it hard for the fear to fade. One of these was the way he handled an incident of late:

> On many occasions, some farmers in the community took their livestock to graze in Kenelm's field without permission, leading to the field being overgrazed. He warned them several times to stop doing so, but they ignored him. He fenced the field, but they broke the fence. He was not strong enough to fight and he didn't want to report the matter to the community leaders, because he felt that they would not listen to him. Suddenly, an alternative solution came to his mind. For some time, he had been collecting straws, plant branches and tree trunks, pigments, wools and animal skins. But no one knew what he was going to use them for. Suddenly, what he gathered them for showed itself. He used the materials to set up an unearthly image which was very terrifying. The image wore a long black dress and a black hat.

With one hand, it held a long stick which was like a shepherd's crook, and with the other hand, it clutched a carved image of a miniature wild cat. It had glaring eyes and clenched teeth. On it's face was a sinister smile. When the wind blew on the image, it swayed in such a way that gave the impression that it was dancing, rejoicing for seeing someone coming close to become its victim.

A farmer, Acton, had taken his sheep to graze in Kenelm's field. Sighting the image, he was so affrighted that he cried out in fear, forcing his sheep to run away with him. But such unusual force, and scream frightened the sheep, leading to their scattering. Consequently, he lost two of the sheep.

The whole community believed that the image had magical power with which it made the sheep to vanish into the land of evil spirits. With this belief, no one in his right sense dared going to Kenelm's field anymore. But it never occurred to them that the sheep actually strayed into the Marshland community, where hungry hunters caught them as a supplement to their game meat.

After the incident, Kenelm's image in Highland was like the image of one who worships evil spirits. It was things like this that reinforced Ward's belief that his family was not in a good relationship with their gods and that the consequence was the infertility of their soil.

*

The winter had arrived in full force, turning the whole landscape into a snowy wilderness. The atmosphere was dull and silent, and ominous cloud hung in the air. Leafless trees stood as though they were mummified. The whole forest was devoid of life.

Highlanders had never experienced such cold and eerie weather before. Ward was growing up and his parents had died. He was lonely. Every night, he heard vibrant voices coming from huts in the community. He knew that families were enjoying their storytelling, but he had no one to share stories with. Added to this, his foodstuff was running out and he feared that it might not last up to the end of the winter. In the middle of this, a new idea came to his mind.

Ward knew that although his soil was not fertile, it yielded something which was a supplement to his forest food. He agreed that if he totally abandoned agriculture, he would not survive. So, he wanted to do something to make his soil fertile. He decided to meet Alger, the Highland priest. He wanted to take game meat to the priest and beg him to sacrifice it on his behalf to their gods as an appeasement, so that they would bless his soil. Also, he wanted to use the opportunity to beg the priest to conjure up his father so that he would hear from him why he had not honoured his promise.

The father made the promise when he was about to pass on. Shortly after Ward's mother died from her illness, Kenelm's health began to deteriorate. Noticing how worried Ward was, he reassured him that their gods would prevent his death in sympathy for the death of

Ward's mother. But shortly after the reassurance, it began to appear that their gods were unsympathetic. Kenelm's health continued to worsen. It was obvious that he was going to die. Then he called Ward and reassured him, promising that if he died that his spirit would help him. Finally, Kenelm died of the illness and up till now, Ward had not received any form of help.

One day, after the winter, Ward went out hunting. He returned with small mammals and birds which he killed. He took them to the priest and made his request.

Alger was a difficult man. He looked at the game animals contemptuously, and turned to Ward and asked him with an evil grin: "Ridiculous game meat as an offering to the immortal? So, you inherited a curse on your father as well as inheriting his behaviour, a behaviour for continuing offending our gods?" He twitched in shock and then bellowed: "Carry this abominable thing and go away before I place more curses on you!"

In panic, Ward quickly took the meat and ran away in distress.

Following the harsh treatment, Ward completely abandoned agriculture and relied on hunting and gathering as his sole means of living. Forests and woods became his real home, giving him the name 'Wild Boy of Highland.' It was this piteous life that the Creator of all things, the merciful one had considered. An intervention was imminent.

Chapter 3

One bright summer day, Ward was hunting. He had killed some animals and was having a rest, dozing under an oak tree. Then he fell into a trance. Following this was a voice calling him: "Ward! Ward! Listen," said the voice. "Your low status and suffering have been noticed. You are now chosen to manage a sacred property. Therefore, leave your father's land and go towards the place where the sun sets. There, you will see a land that is a symbol of nature's glorious order. From its forest, you will get effective weapons against all diseases. Its animals will be your game meat. Through your good management of the land, its soil will always remain fertile and you will be a rich farmer.

"Although I give you the land and everything in it, never forget that they are my property. I only lend them to you. All living things in the land, whether big or small, treat them with care. Don't destroy them. Use them wisely. This is the greatest law you must obey in the land. Through obeying the law, you will prosper and develop deep insight into hidden things; you will be a blessing to people around you; you will relay enlightening messages to all mankind, and be called a revealer of the mysteries of nature. Remember, the root of your success, peace and happiness will be your adherence to this instruction, and your misery and sorrow will arise from your negligence of it."

The spell was broken and the die was cast. Ward woke up from the trance and like a darkness that vanished in the presence of bright light, his tormenting belief, fear and anxiety were replaced by new hope. He was invigorated. He thought deeply about the matter and then soliloquised delightfully: "Truly, this is the voice of the merciful Creator." He got up and hurried to his home. When he arrived, he prepared his food and put it in his bag. With the food, his dog, his bow and his quiver which was full of arrows, he set off. By now, he was heading towards Marshland, a community near Highland. The people of Highland believed that the inhabitants of Marshland were rejected by gods. The reason behind this was obvious. Marshland was a marginal land. Its soil was more infertile than Highland soil. The way of life of the inhabitants of Marshland was like life before agriculture. The people were not farmers and did not have cult of fertility deity. Hunting and gathering were their means of living, and they believed in magic and animal spirit. Expression of the belief was shown in their adornment of their weapons and tools with engraved representations of their game animals and their possession of sculptured models of the game animals. As a magical preparation, prior to hunting expedition, they performed invocatory dance or mock fight before their objects. So Highland people detested Marshland community and felt that anyone who chose to live there must be a cursed one.

Alger, the priest saw Ward when he was heading towards the Marshland community. Before, he suspected that Ward lived in an unknown community from which he went to the forest to hunt. This was because since Ward totally abandoned agriculture, he was rarely seen in Highland. Now that the priest saw him heading towards Marshland, he felt that the community in which he lived was Marshland.

"Oh," the priest murmured in horrified mood, "being unable to bear retribution anymore, he left and made forests and Marshland his home, living with people like him. Our land will now be free of desecration and our gods will weep no more."

Seeing Ward's home as a desecrated place, the priest went there and carried out ritual purification.

Ward had crossed Marshland and was heading towards the horizon. He made a very long journey and finally arrived at his destination. He didn't need to be told that he was at the land. The sight, sound and smell of the environment made it clear to him. It was a land of abundant animals. Its grassland was a luxuriant meadow, and its forest, lush vegetation full of diverse fruits and seeds. The air was filled with the fragrance of colourful flowers, and voices of animals were like a therapeutic sound. Ward was uplifted. "No doubt, this is the place," he intoned in affirmation. "I have been taken away from the valley of darkness and sorrow, and planted in the valley of bright light and hope, a valley of flora that revives and sustains."

Ward named the land Floradale. Since then, it has been known by the name even to this day.

Ward settled in a big cave in the land. It was from there that he went hunting; gathering fruits and seeds. His hunting and gathering were not just for getting his food, but also as an opportunity to explore the forest.

One day he came to a hill, and from there he looked at a distant place where faint vegetation seemed to stand on top of a hill. He went towards that place. He had covered a long distance and was at a vast grassland with sparse trees and shrubs. As he travelled, the faint vegetation continued to increase. He could see dim smoke rising from the vegetation.

"It seems a sign of human activity," he thought. The more he moved on, the clearer the smoke became as it rose to the blue sky. He had now come to a place where the vegetation developed into thick forests and woods which curtained off the smoke. As he walked, he arrived at an area where farmers planted crops. It was there that he started hearing a faint crowing of cocks and barking of dogs. He moved on and finally came to a place where he saw people, their huts and domestic animals. It was Oakland community. He then went to their market and saw different commodities – grains, livestock, game meat, agricultural tools, and hunting weapons.

In the market was Edgar. He had what Ward wanted most. Ward greeted him, drawing his attention. "What do you want in exchange for these?" he asked, pointing at some mattocks, hoes and sickles there.

"What do you have with you?" Edgar replied. He suspected that Ward was a hunter because he was with his dog, bow and quiver. "You are a hunter?" he went on. "I want to exchange them for good game meat. Can you afford that?"

"Yes," replied Ward. "I can, because hunting is my occupation."

"Then where is your game meat for the exchange?" asked Edgar.

"I don't have it with me," said Ward. "I will bring it when I come again. I come from a long way to this place just to find out if this market has what I need. My aim is to find out first, and then come back later for the purchase."

"Where do you come from?" Edgar asked.

"Floradale," Ward answered.

"Floradale," Edgar wondered. "This sounds unfamiliar. Could it be one of those hamlets or clans found in the hinterland region?"

"No," replied Ward. "It is not in the region you are talking of. It is a unique land which is between the region and your territory, and whose inhabitants are its animals and plants, and I and my dog."

"Oh!" Edgar said in surprise, "you must be a devoted worshiper of the god of the forest. You left your people and settled in the land so as to be closer to the god, serving him fully and earning the reward." For a moment, he gazed at Ward, smiling suspiciously, his eyes twinkling in excitement. "See the tools and show me the ones you want," he continued. "I will keep them for you, but first bring to me what the god offers you."

Quickly, Ward laid down his bow and quiver, and examined the tools. "They are good, but I would like better ones," he said.

"I have those ones at home," Replied Edgar.

"I would like to see them please," said Ward.

"Wait a moment," Edgar requested. He communicated with his nearby market friends. Having left his stall for them to look after, he walked to his home with Ward. As they walked, the memory of the smoke Ward saw during his journey came to his mind.

"On my way to this place, I saw a thick smoke coming from this region," Ward said to Edgar. "I thought that every place here was on fire. But when I arrived, the smoke seemed to have vanished. Where is it coming from and what is the cause?"

"Oh dear!" Edgar said in surprise. "As a hunter you should know where the smoke is coming from and why. You are indeed a stranger and very young. The smoke is from burning forest far away from our community. If you are at a distant place like the hinterland or Moorland community, the smoke gives the impression that our community is on fire. Our society consists of farmers and hunters. The burning of the forest is the activity of our hunters. It is a practice which started from time immemorial. In the past, hunting and gathering were our predecessors' means of living. To improve their mobility and reduce escape cover for their game animals, they adopted the practice. Also, the practice is a means of improving the amount and quality of grazing for the wild animals. Today, we still have people who practise this way of life. They do a little farming and so live mainly on forest food."

"Why do they do this?" Ward asked suspiciously. "Did something happen to their soil?"

"No," replied Edgar. "Their soil is fertile. They just chose that way of life. They prefer forest food to agricultural produce and worship the god of the forest more than they worship the fertility goddess."

"And their soil still remains fertile; the goddess continues to bless their soil?" Ward asked wondering.

"Oh!" Edgar replied in amusement. "The mother goddess is very forgiving, merciful and caring. She sees us as her children. A mother doesn't punish her children because they are ignorant or weak in appeasing her. She may chastise them for their wrong doing, but not punish them fatally. If she does, she will lose them. The goddess only punishes one severely if the one does something that is life-threatening."

As Ward conversed with Edgar, he learnt about the people's culture. He was impressed by the good relationship between the people and their deities, and their warm acceptance of a visitor.

When they arrived, Edgar's success was obvious in his homestead. There were many livestock there, and at one side of the homestead was a big barn with plenty of crops. At the other side was a smithy where Edgar made most of the tools he traded. Attached to the smithy was a store for agricultural tools and hunting weapons. He had a big hut at the centre of the homestead. It was there that he received visitors. Behind the hut was a shrine which housed images of deities. It was there that Edgar offered sacrifices to their gods and prayed regularly to them for the welfare of his family, community and his success in his occupation.

Edgar took ward to the store and showed him the tools in it. Ward examined them and showed him the ones he wanted. In addition to the agricultural tools he wanted, he also selected some high quality hunting weapons. They then went to the reception hut and discussed details.

"I have fertile soil," Ward said. "But I have no tools to till it, and no crop to sow in it. As I said earlier, I come to this place in search of these items. Now that you have them, and in addition you have the type of tools I need for hunting, I wish to come back with some game animals for the purchase. But what type of game animals do you want me to bring for the exchange?"

Edgar nodded in satisfaction and then replied: "I am delighted by your ambition. I helped many young people like you in the past, and I am happy to see the progress they are making today. I helped them because of their wisdom and determination. Yes, it is good to be a hunter, but better to be a farmer. Earlier, our predecessors were hunters and gatherers, but later they adopted farming. They did this because agricultural occupation is a way of making a good living as well as reinforcing good relationship with the gods. It is a sacred occupation where one imitates the work of god. It is a profession through which one contributes to maintenance of life. Pursue the career and I am here to support you. But first bring the game animals you have. Whether they are big or small, there will be crops and tools which will be the equivalent."

Ward was very happy. He thanked Edgar with all his heart and promised to be supplying him the game animals. He then returned to Floradale.

A link was now formed as a foundation for fulfilling the purpose for which Ward came to Floradale.

Chapter 4

By supplying game animals to Edgar in exchange for crops and agricultural tools, Ward established agricultural occupation. At the early stage, Edgar lent him his cart for transport, making it easy for him to transport his crops, livestock and game animals to Oakland for trading. Ward employed people from Oakland to work with him on his farm. Edgar made him aware of Owldown market. The market was the biggest in the whole territory. It was from there that Ward purchased draft animals. This improved his work in terms of moving heavier loads and ploughing his field effectively.

People from different communities came to Ward for his agricultural produce. One day, a weary-looking group came to him. After Ward welcomed them, Hunter, the leader of the group spoke: "We come for your help," he began. "We are from Woodland. Our occupation is hunting and gathering. This is because we can't afford the things that are required for farming. It is only recently that we have been hearing that our neighbouring community, Brookdale hamlet is doing well in farming. We investigated and found out that this is true. We heard that it was you who made this possible by providing the people what they needed to start the occupation. These people are those we used to laugh at for preferring farming to hunting and gathering. We laughed at them because they killed wild animals and exchanged them for meagre crop. With poor tools they grew the crop in pain. In the end, they got poor harvest of which the quantity was less than the quantity of the forest food they would have gathered within the time it took them to do the planting. But these days we are astonished by their success. They have barns full of crops and livestock.

"We come to beg you to be merciful and lighten our burden. We beg you to help us as you helped the Brookdale people." He laid down before Ward small mammals and birds they had killed. He then continued. "Throughout today, we have been hunting. These game animals before you are all that we were able to obtain. We beg you to consider our poor status and give us your crop in exchange for these."

What he presented before Ward was laughable, more ridiculous than the one Ward presented to the priest of Highland as an offering to their gods. The hunters' appearance showed the extent of their hard life. Ward looked at the youngest one in the group and noticed how weak and malnourished he was.

"Do you go hunting with this child?" Ward asked.

"Yes, we do," replied Hunter.

"What is his name?" asked Ward.

"His name is Stan," Hunter replied.

"How does he hunt and gather with you in this his state?" Ward asked in disbelief.

"Although we go with him to gather forest food, he doesn't play any useful role," replied Hunter. "Considering his circumstance, we take him with us only to share with us whatever we obtain from the forest. He is an orphan. In obedience to the law of our ancestors, we have to help him as much as we can. The law states that any orphan, widow, or disabled one in our clan, must be looked after by us. If the law is disobeyed, what follows is unbearable retribution. His accompanying us to the forest is not just for sharing the forest food with us, but also for reducing his loneliness."

Ward's mood suddenly changed. He had remembered his past and the difficult life which led to his visiting Alger, the Highland priest, and how the man treated him. Tears of compassion ran from his eyes. He replied to the men as he wiped away the tears:

> "I have heard you," he said. "I will look into your problem. Give me time to think about it. Go home. Take the game animals with you. You have to come back later. By that time, I will have an answer for you.

"You said that one of the ways you reduce Stan's loneliness is by ensuring that he accompanies you. To show me that you are telling the truth, let me see him accompanying you when you come back."

Ward told them the day they had to come back. They thanked him and promised to come with Stan. Before they went home, Ward gave them some food.

After the people had left, Ward thought carefully about their needs. He knew that in addition to the crop which the men needed, they would also need good tools. If he gave them the crop without the tools, they would use poor tools to plant it, and that would be a waste of time. So, he took some of his livestock to Edgar and exchanged them for good hoes, mattocks and sickles.

By the time the men came, the tools and many bags of grain were there waiting for them. Ward then addressed them.

"These bags of grain, and these tools are yours," he said. "I don't want you to give me anything in return. All that I ask you is to start agriculture with them. In your own words, you said that you need these to start agriculture. I am sure you will do what you said because you have already proved to me that you are trustworthy. Last time you promised that you would come back with Stan. You fulfilled the promise.

"I am concerned about Stan's poor state of health. He can't farm with you in this his health condition. Take the grains and the tools with you, but leave him to stay with me until he is fit enough to work with you."

The men were overjoyed. They thanked Ward again and again and agreed to leave Stan with him. Ward then entertained them with food and drink. After the entertainment, he asked his workmen to help the hunters to transport the grains and tools to the hunters' home. As the men went home, they chanted ecstatically, just as they usually did when they were lucky enough to kill a big wild animal.

Ward took good care of Stan. With a mixture of nutritious food and herbal medicine, he nurtured him into a vigorous and healthy youth. Through learning from Ward, Stan developed skills in farming. Finally, Ward took him back to his clan where he joined his people in farming.

The hunters had become so successful in farming that they traded agricultural produce in communities, including Oakland. Knowing who they were before and seeing their success, the people of Oakland were so astonished that they agreed that Ward was a blessing to people around him. Yet there was a greater thing which Ward did in Oakland and which boosted his image more.

One day, Ward visited Edgar and saw him in a sad mood. "What is wrong?" Ward asked.

"We are about to face unavoidable bereavement," Edgar replied sadly. "Blake whose father died recently is seriously ill. We are worried, not just because of his ill-health, but also because of what will happen to his ageing mother if he dies. He is the only one who looks after the mother. We tried on him every herbal medicine and carried out ritual healing, but all failed. Every community has shrines where its people offer sacrifices to their gods, praying for them to solve their problem which is beyond their solution. But when the problem remains unsolved, the final place to go to is Owldown. It is there that you will find the best healers. We appeased our gods and prayed for their intervention, but the problem was not solved. So, I and Woodrow went to Owldown and brought their herbalists. But their effort failed. They suggested that we seek the help of their priests, and we did. The priests carried out ritual healing but failed. We enquired from their oracle if we have offended the gods. He carried out divination, but failed to find an answer. We then consulted their astrologers. They observed the movement and relative position of the heavenly bodies and could not discern any change that explained the cause of our problem. So, it became obvious that we have no choice but to come home and wait for the worst."

"Where is the patient?" Ward asked.

"Come, let me show you," said Edgar. He and Ward went to see Blake.

After examining Blake, Ward went to Floradale and returned with some medicinal herbs and started treating him. It didn't take much time before the treatment started yielding a positive result. The patient who could not stand up was now walking slowly. As Ward maintained the treatment, Blake fully recovered. Ward took him to Floradale and employed him as one of his workers. With the occupation, Blake took better care of his mother.

Oakland was highly amazed. Ward had brought honour to them by providing that what Owldown failed to achieve, Oakland now had someone who could achieve that. The people were very proud of Ward. Shortly after this, they organised a festival in the community. The aim was to bring communities to Oakland and make them aware that Ward had done what Owldown failed to do.

In the festival, many people from different communities were invited. Ward was there as a celebrity and Oakland musicians performed, chanted, singing in praise of Ward:

"Thanks to the god
The god of the gods
He reared his child
Crowned him with gifts
Made him our man
The son of our soil
A blessing to us."

It was this image of Ward that attracted an unexpected girl to him for marriage. The girl, Willow was from a noble family in Oakland. Her father, Woodrow was very influential. He represented Oakland in meetings held in Owldown. In the meetings, emissaries from different communities gathered to discuss how to continue preserving and strengthening social order in the whole territory. Owldown people respected Woodrow and had a good relationship with him because of his wisdom and the useful ideas he contributed during the meetings.

Willow was very beautiful. Suitors from noble families had been coming from Owldown, proposing an engagement with the girl. Woodrow was very happy because her marriage with a son of Owldown would strengthen his relationship with the community. But Willow found Ward more attractive than any of those men. She had been looking for an opportunity to become closer to him. Willow was a friend of Edgar's daughter, Bridget, and they regularly visited each other. As Ward was always in Edgar's place, this made Willow's visits to her friend more frequent. It was during the visits that she developed friendship with Ward. The relationship continued to grow stronger and stronger until it became obvious that it was leading to a marriage.

But there was a period when it seemed that the relationship between the two was fading. It was one of those periods when Ward was very busy, meeting increasing demand for his farm produce, game meat and herbal medicine. For some time, Willow did not see him in Oakland. She suspected that Ward had found a new lover, probably from Owldown. She took her family's horse and rode to Floradale. Ward was about to set off to Owldown to supply livestock to his customers when he saw her on the horse, speeding towards him. He stood, looking at her in surprise. Having approached him, she alighted from the horse and with a sigh of relief threw herself at Ward, her hands wrapping around his neck.

"I missed you so much," she said affectionately.

"But I missed you more than you missed me, oh goddess of beauty," Ward replied as his hands wrapped around her slim waist. "I love you so much," he added.

In response, she leaned in for a kiss. He could not resist her luscious lips. They kissed and hugged each other. Her breasts like newly ripened fruits pressed on Ward as his hands ran up and down her 'S' shaped body, feeling her curves. It was not just a feeling of woman's attractive body, but also a feeling of her good health and maturity for childbearing.

The die was cast. Whatever business Ward had at hand was suspended. That day, they went to Oakland to announce their engagement. This was later followed by their marriage. Within a year after the marriage, Willow gave birth to a bouncing baby boy. He was named Edmund. Ward's household was now a noble family, well respected by Oakland and the neighbouring communities.

Ward was a man on a course for fulfilling a divine task. So far, it was obvious that he was achieving the goal, and the key to the achievement was his good relationship with all life around him. However, time will tell if he was going to forget the key and face difficulties.

Chapter 5

In the beginning, the thunder lightning took the role of maintaining the soil fertility. It struck nitrogen gas in the atmosphere, turning it into a form that finally became a plant nutrient in the soil. Some soil organisms were aware of this and so they made a request to plants:

> "We are those known as rhizobia," the organisms said to the plants. "We know that the thunder lightning makes a nutrient which you absorb with your roots. We can make the nutrient as well but in our own way. Can we enter into a relationship where we will be supplying you the nutrient as a supplement to the one the thunder lightning provides for you, and in return, you supply us the food that gives energy?"

The plants were very happy. They accepted the request.

"But where are you going to make the nutrient?" asked the plants.

"There are some of you here who are called legumes," the rhizobia replied. "If they can allow us to build nodules on their roots, that will help. The nodules will be our home as well as our factory in which we will be making the nutrient." The legumes were happy and enthused. They agreed to the request. They promised that from that day onwards, wherever they were, they would continue reminding rhizobia around them that their roots were available for them to build the nodules. The rhizobia rejoiced and built the nodules, making and supplying the nutrient to the legumes as the legumes supplied them energy-giving food in return.

Other soil organisms noticed the mutualistic relationship between the plants and the rhizobia. They called the plants and made their own request.

"We beg you to allow us to be part of your community," the organisms said. "We are the organisms known as decomposers. We are called decomposers because we decompose or break down dead things. What we feed on include plant litter, animal remains and waste. During the feeding process, we break down the materials and release plant nutrients in them. If we live together with you as one community, we will feed on your litter, releasing the nutrients, making them available for you to take in."

The plants rejoiced and thanked the decomposers. "Yes, we will be happy for you to be members of our community," the plants said. "Your feeding on our litter will help us a lot. It will increase availability of nutrients for us. It will also prevent the litter from accumulating to the extent that will make us uncomfortable."

Therefore, the two parties became good neighbours, each benefiting from one another. Now, a mutualistic relationship between plants and soil organisms was completely established, and so the soil became a living system.

Ward's method of agriculture was in harmony with the well-being and existence of the organisms. Regularly, he manured his field with farm waste and litter, making them available for the decomposers to feed on and release nutrients in them, thereby adding more nutrients in the soil for his crops. He grew legumes and then harvested them, kept some as his foodstuff and put some in the soil for the decomposers to release nutrients in them for his crop.

The life and well-being of animals are linked to the plant nutrients. In general, the nutrients ensure plant well-being and growth. When animals including man consume the plants, they gain the nutrients for their own well-being. One of the nutrients is the one which the thunder lightning and rhizobia supply. Plants use this nutrient to make protein and when animals including man eat the plants, they gain the protein; and they gain more of the protein when they eat the legumes. So, there was no doubt that Ward's method of agriculture benefits his crops, and the soil organisms as well as benefiting him.

Some time ago, Ward noticed some pests in his field crop. He wanted to do something about it, but he had no idea of what to do. So, he went to Oakland to make enquiries. He had arrived at his in-law's place. He met Brent, his wife and their children. They were bringing out crops to sow in their cultivated field. It was not an environment in which Ward wanted to say what he came for. First, he greeted them and then spoke to Brent.

"Can I have a word with you?" he asked, walking away from the group. Brent followed him. They arrived at a spot where Brent's family could not hear them.

"I feel embarrassed to make this enquiry, but I have to," Ward said to Brent in a low voice. "But before I go on, I would like your father to be present."

"Let's go to him," Brent suggested and led the way to Woodrow's hut. When they arrived, Woodrow was not there. Brent asked Ward to sit down and wait. He then went straight to a shrine in the homestead. He peeped through a fence surrounding the shrine before returning to Ward.

"Let's wait," Brent said to Ward. "He is doing something very important. Once he finished it, he will come."

Woodrow was in the shrine carrying out the sacred duty of the season. He was offering a sacrifice to the fertility goddess. The deity had to be appeased to bless the soil and the crops that were being sown in that period.

After a while, Woodrow walked in.

"Oh, is that Ward?" he asked in surprise. "What brings you here?"

"A problem," replied Ward.

"What problem is that?" Woodrow asked as he sat down.

"It is a pest problem," Ward replied.

"Oh, pests and their problem," Woodrow said incredulously. "Have they arrived in Floradale?"

"They have", Ward began. "For some time, I have been noticing the insects in my field crop. Today they seem to vanish, the next day they re-appear. Although the population is very low and they haven't caused any significant damage to my crop, I fear their presence. I feel one day they will multiply and destroy my crop. I want to do something about it before it is too late. But I don't know where to start. This is why I come to seek your help."

As Ward spoke, Brent laughed in a way that expressed that Ward was enquiring from the wrong person.

"It is good that Brent is here," Woodrow said. "He can tell you how insect pests invaded our crops on many occasions, leaving nothing for us to harvest. We tried everything we could to control them but failed. All that is left for us to do is to keep on praying for our gods to help us."

Suddenly, Woodrow stiffened in mixed surprise and suspicion. "But if I may ask," he continued, "why do you come to us for a solution to a problem like this? You have more knowledge than any wise man among us. I am surprised to see you, the master of agriculture, coming here to enquire from us how to solve agricultural problem. If another person had come here to make this enquiry, I would have advised him to see you for the solution." He paused briefly, looking at Ward with facial expression that seemed to suggest that Ward was joking. "Anyway," he continued, "I don't have the answer. If I do, I would have solved my own pest problem. The only thing I can do is to pray to the goddess not to allow an occurrence of the problem you are anticipating, just as I pray for her to stop my already occurring one."

"I don't know how to tackle this problem," Ward persisted. "If I do, I wouldn't be here. Tell me, is there anyone you know who has been in contact with farmers in distant lands? I might gain a clue from such a person, because it is said: 'A traveller is wiser than a person of grey hair.' He might have gained useful knowledge from the farmers or heard a helpful story from them."

"In that case, the person you need is Straw," Woodrow replied. "He is a well-travelled man and the oldest one in Oakland. He has been in contact with small and big societies far away. He has stories about their cultures. Moreover, he is the master of our oral tradition, myths and tales."

"Please can I see him," Ward requested.

"Brent, take him to Straw," Woodrow told the son.

Quickly, Brent and Ward went to the man. When they arrived, Ward witnessed the truth about Straw. The man's age, experience and status were clearly visible. He was a tall man with a stoop, and he had a bald head and a bushy white beard. When he walked, he supported himself with a walking stick.

Straw had a large extended family. His homestead was so huge that it was like a small community in Oakland. At the centre of the homestead was a big hut in which the old man lived. Within the hut, one could see the evidence of the extent of the man's contact with people in faraway societies and how long that was. In it hung works of art of various tribes – sculptures, hunting tools with engravings on them, skulls and horns of exotic animals. The original colours of these had completely gone, the major cause being slow fading and long-time exposure to smoke from a log fire. Although all Straw's sons had attained their manhood and had children and grandchildren, he remained the head of all the families and was treated with great respect by them. The position of his hut in the homestead was an indication of this. His sons regarded him as an intermediary between them and their ancestors.

Brent had introduced Ward to Straw after they came into his hut. Ward narrated his problem to the man and made his enquiry.

Straw had meditated over the matter, and after a short silence he began: "My son, I know the cultures of many tribes I have been to in the past. I remember their method of agriculture. The method of agricultural practice which was handed over to us by our predecessors is very clear to me. Out of my knowledge, only one thing came as a solution to your problem."

Ward waited expectantly.

"The solution is appeasement of the fertility goddess," Straw declared, looking at a shrine opposite his hut as though the goddess was there listening. A wave of disappointment and dismay swept over Ward.

"Pests are always present in field crops," Straw continued. "This is normal. We don't worry about it.

We only become concerned when the population increases above its usual level. This is the time they cause significant damage to crops. One of the reasons we offer sacrifices to the goddess is to appease her to prevent this. She accepts our sacrifices because the pest population rarely reaches the worrying level. Even when it does, the crops are not totally damaged, we still harvest something that sustains us. Go home and offer a good sacrifice to the goddess. Pray to her not to allow the pest population to increase above its usual level. If you heed my advice, your crop will be free of pest destruction, unless you are one of those exceptionally unfortunate individuals. These are the people whose relationship with the goddess has become so bad that a sacrifice can't repair it. If you are one of such people, the best advice I will give you is to stop farming and choose another career. From a story I am going to tell you, you can derive a reason for my advice." He paused to remember the details of the story. Ward waited in unease as the memory of his circumstance in Highland was refreshed.

"Following a repeated profane behaviour of a family in a land called Winterfield," Straw began the story with a voice from which one could easily discern the extent of tragic the story was, "the family was cursed by the goddess and any soil any of the members of the family stepped on became infertile. The curse was inherited by their descendants. In the end, the descendants abandoned agriculture and resorted to hunting and gathering. But one of these people left their community and settled in Donnockforest. Later on, he went to the Owldown priests and requested for the curse to be nullified. His request was granted. He started agriculture and at the beginning it seemed that all was going well. But suddenly, the ill-luck returned to him in full force. His soil became very infertile and any crop in his field was invaded by pests. The man returned to Owldown priests for their help. The priests appeased the goddess and entreated for clemency. Despite all these, they failed to achieve anything.

"In the end, the man returned to hunting and gathering and spent the rest of his life in that career," he paused briefly and continued with an instructive tone. "If you look back to your past and see that your predecessors were in good relationship with the goddess, then go and do what I told you and preserve the relationship. If you observe otherwise, then my warning to you is this: continue farming but at the same time be prepared to spend the latter part of your life in a non-agricultural career. No any sacrifice can solve the problem. Something other than that is needed. But who knows what that is?" He looked at Ward momentarily with a serious face and answered, "No one. Even Heartroot with all his sacred knowledge doesn't know it. It is unknown to the Owldown oracle, and remains a great mystery which the incantation of the faraway wizards can't unlock." He highlighted the truth of his assertion by nodding repeatedly and confidently.

Noticing Ward's depressed mood, Brent spoke to him, trying to reassure him. "Ward," he called, "the pests in your field are not too much for you to worry about. There are more of them in our field crop and we don't worry about it. You are the luckiest one among us, and the most favoured by the gods. Go and continue your farming. There is nothing to worry about."

After listening, Ward thanked Straw, pretending to be satisfied with his instruction. He then left with Brent to tell Woodrow what the old man told them. Woodrow reassured Ward just as Brent did. Ward then returned to Floradale in fear and frustration.

When Ward returned home, he sat down and thought deeply about the matter. What worried him was not the failure to find a realistic solution to his problem. It was a new thought which he developed following Straw's story and advice. He thought about the background of the man in the story; about the extent he went for nullification of the curse and about his failure in the end. He compared himself to the man and saw a clear similarity. He was deeply perturbed.

"Am I still unsafe!" he lamented. "Is the fertility goddess waiting for a time to curse my soil to be infertile, and send pests to destroy my crops? What is this that I heard from this man?" Ward was frightened by the story. He hated the teller and vowed never to ask anyone anymore how to solve the pest problem. He had decided to work out the solution by himself.

Chapter 6

For some time, Ward had been thinking about what to do to solve the pest problem. Suddenly, he remembered some plants in Floradale whose extracts were very toxic to insect pests. He wanted to try the natural pesticide on the pests. Then, it was like two voices speaking to him. The first voice began:

> "Ward," the voice called, "you went to Woodrow and Straw for information about how to solve the pest problem. Although you didn't get the information, there is a useful lesson you learnt from them. The lesson is that, in every part of the world you can think of, people have their culture which they practise, and religion is part of the culture. The practice of their religion seems to be a way of regulating their behaviour so that they live and function in harmony with nature. If a problem beyond their understanding and solution arises, they don't tackle it by themselves, for fear of making a mistake. Instead, they approach their gods through prayer and sacrifice for explanation and the solution.
>
> "You heard from Woodrow and Straw that a field crop is never free of pests. Woodrow told you that sometimes pests damage their crop, leaving nothing for them to harvest. But he didn't tell you that their farming career

ended due to the pest infestation. Could it be that the gods allow pests to exist in the field crop for a hidden reason, but never allow them to destroy the crop completely? Just like the voice that spoke to you in the past, could it be that the people's gods also spoke to them? The voice that spoke to you told you to go to Floradale and become a great man, and in return you should take care of the organisms in the land, whether they are big or small. Could it be that these people's gods spoke to them, telling them to do the same thing you were told to do? Could it be that their religious practice guides them to behave in a way that ensures that they don't destroy the organisms?

"So, make sure that what you call pests are not part of the organisms you were told not to kill. Therefore, study the organisms in your field crop and become aware of their importance, and then know what needs to be removed and what needs to be kept."

Like one who aimed to solve a problem through war, but suddenly found a peaceful approach as an alternative, Ward thought deeply about the words and was about to change his mind, and take time to study the roles of the organisms in his field crop. Then, the second voice interfered:

> "Ward, don't listen to that unwise advice," said the second voice. "When did you start wasting time, studying an illness which you already know? Don't you know that doing so, means giving the illness time to kill the one who is suffering from it? Blake was ill and despite their religious

practices, Owldown and Oakland failed to heal him. But when you were called, you went and healed him. Did you heal him after studying the illness?

"You enquired from Woodrow and Straw about how to solve the pest problem. Judging from what they told you, they don't know the solution despite their religious practices. They don't know, not because their gods don't want them to know. It is because the gods themselves don't know the solution. Go and prove to them and their gods that you know better than them.

"As a medicine solves health problem, so does a pesticide solves a pest problem. Your crop is part of the life which you are here to protect. Pest is a disease against the crop. Pesticide in the forest is a weapon waiting for you to use and save the crop. But you are being advised to give the pest time to kill your crop, by taking time to study it. Save your crop before it is too late, and prove to Oakland and Owldown that you are greater than their gods."

Ward was relieved and invigorated by this advice. He felt that, not only that a solution of the pest problem was imminent, but also that the solution would change his image to the image of a god. He wished not to hear any other advice other than this one. Since then, he only listened to stories that justified the use of the pesticide on the pest. One of the stories was the one his wife told the family one night.

*

It was winter season. Plants wore snow as their white apparel, and snow covered the earth like sand dunes. Though cold, the world was brilliantly scenic.

The season was a period when Ward had a break from his agricultural work. During the period, he spent more time with his family members – Willow; Tiller, the second son; Lina, the daughter; Edmund and his wife, Mildred; and the couple's son and daughter, Dale and Edith. Sometimes, in the day, the whole family played together in the snow as part of their recreation. In the night, they entertained themselves with storytelling in their cozy hut. Sometimes, a view of the outside scenery made the night atmosphere more pleasant and spectacular. This was when uncountable stars dotted the sky, as the moon glowed, illuminating objects, and the snow reflected the moonlight, enhancing the illumination.

This night, Willow had a story to tell. But it was a different night. The moon and the stars didn't appear. It was as if the dark sky and the earth merged, and became a dense forest of darkness whose undergrowth was all objects on the earth.

"Once upon a time", she began, "all insects had a meeting to discuss an important matter. Before then they all lived on fruits and seeds only. As their population increased, the food could no longer sustain them. In the meeting, they wanted to agree to eat other parts of the plants, in addition to the fruits and seeds. Eating the other parts of the plants would reduce their demand for the fruits and seeds.

MAN AT NATURE'S PINNACLE

"But aphids were more cunning than any other insect. They met together and whispered: 'Let's take from the insects the larger and best parts of the plants – the leaves and shoots. Let's suggest to them that we should sort ourselves into groups, and each group should choose a specific part of the plants and specialise in feeding on it. Let's say that the plant parts should be leaves, shoots, roots, flowers, wood bark, and dead leaves. Let's ask the bees to form their group first, and make their choice. As we are all aware, although they eat fruits, they like to suck flower nectar more than any other food. They will therefore accept our suggestion and choose the flowers. Once they accept the suggestion and make their choice, to pay us back, they will definitely present us before the other insects as the next to make their choice'.

"The aphids then presented their suggestion to the insects.

"'Brilliant idea!' the bees' speaker shouted delightedly. 'Do you agree to the suggestion?'

"'We agree!' All the insects shouted happily. 'Let's then sort ourselves into groups,' said the bees' speaker. 'Which of you want to be part of our group?' He asked. 'Choose the ones you like,' the aphids replied knowing that the bees would choose the butterflies because of their beauty and gentleness. The other insects joined the aphids, asking the bees to make their choice. The bees chose the butterflies. As the butterflies liked flowers, they and the bees had the flowers.

"Favoured by the aphids' suggestion, the bees asked the aphids to choose the insects they like and form their group.

"'We are too many,' the aphids replied. 'If our number is added to the number of our children at home, we will be more than a group. We choose the leaves and the shoots and beg you to leave us as a group. Although we may allow you to occasionally feed on the plant parts with us, know that they belong to us.'

'"That is understandable,' the bees' speaker said. 'Do you agree or not?' he asked.

"'Aye!' the bees and butterflies shouted, intimidating the other insects to agree. All the insects then agreed. It was ants' turn to form their own group and make their choice. The ants' speaker got up and addressed the insects: 'Thank you fellow creatures,' he began solemnly. 'We are pleased with the issue raised and the suggestion made. You know that just like the bees, we have a community and a queen. But the bees have the freedom to make a decision without consulting their queen. Unfortunately, we don't have such freedom. Whatever we do, where we go, and what we eat are authorised by our queen. We are bound by law not to do anything without her approval. We beg you to give us time to hear from her.' The insects granted the ant's request.

"Meanwhile, the insects that had formed their groups and made their choice were allowed to feed on the plant parts they chose. Those that hadn't made their choice were left to feed on the unchosen parts.

"The ants went home and reported the matter to their queen. The queen was alarmed. First, she thanked them for not making their choice and warned against the danger behind such a plan. 'Don't you know that such a plan can lead to extinction of life on the earth?' she asked in horrified mood. 'The fruits and seeds of the plants are the things that sustain us. We are able to get them because the plants produce them. If the insects turn their insatiable appetite against the plants and kill them, what will give us our food? Can you go out gathering and ever return with anything anymore? Beware of these insects. They are stupid and dangerous. Apart from the bees, which of them is capable of thinking properly and behaving wisely? Do they have an organised community and a leader like you do? Where is their thought about the future and their preparation for it? We laugh at them in their ability to breed prolifically but inability to care for their offspring, but I must warn you that their behaviour and lifestyle are no longer something to laugh at and ignore. They are dangerous. Even the bees, despite their wisdom, fail to comprehend this. Now listen, all the time they and their uncountable offspring have been consuming the seeds and fruits in the forest, almost leaving nothing for us to gather, did we complain? Did we not ignore them and with fortitude travelled far and wide in search of anything they missed out? Did we not live in our kingdom with no resentment against them? It seems they are not satisfied with their crime and so decided to commit more by going to destroy the plants upon which our life depends. When they gathered there with you, it was not for saving life, rather it was for destroying it. This is their aim. You must not allow this to happen. You have to start a training to prevent this.'

"The queen started training the ants as fighting solders. As the training was going on, they sent their messenger to inform the insects that they had a problem in their community which they hadn't finished solving, and so requested that the insects give them more time to finish solving the problem, and that once the problem was over, they would let them know. The insects granted them the request. Meanwhile, the insects kept on eating the plant parts, and as they did, their mouth parts continued to develop, adapting to the diet. But the ants continued sharpening their teeth, developing their jaws for killing the insects.

"After the training, the ants' queen gave her order: 'From today onwards,' she said, 'as the insects chose the plants as their food, so have we chosen them as our food. The more they grow in number, the more our appetite for them as our food grows. As they work hard, eating the plant parts, so should you work hard in killing them, bringing their bodies to our community.'

"The ants then sent their messenger to inform the insects that their problem was over and that they were ready to meet the insects for forming their group and choosing the plant part they preferred as their food. The messenger went and gathered all the insects together and delivered his message. 'Fellow creatures', he addressed them, 'we have completed the solution to our problem. Our queen is pleased with your plan and agreed to it. She apologised for the delay. The delay is because we have been renovating our homestead. This involved chopping off trees with our teeth. You can see how big my jaw is. As we carried out the activity, our jaws continued to develop, adapting to the job. Our queen is so pleased with your plan that she has some gifts for you. We will bring the gifts in the next meeting. Just tell us when you want the next meeting to take place and we will come to form our group and choose our own plant part. Encourage everyone to attend the meeting because we will bring a huge amount of gifts for you.'

"All the insects rejoiced after the message. They thanked the ants' messenger and promised to come to the meeting with all their friends and relatives. They fixed a date for the meeting. Then the ants' messenger returned home.

"The day for the meeting came and all the insects came in large numbers. Like an invading army, the ants marched out and fell upon them, slaughtering them and carrying their bodies to their community. In confusion, the insects that survived the onslaught scattered into every part of the earth. But the ants pursued them anywhere they ran to. Wherever they found the insects, they set up a new community there with a new queen, killing the insects, taking their bodies to their community. But the insects had already adapted to eating the plant parts and could not change. The ants therefore continue to kill them even to this day. This is why you always see the ants carrying insect bodies, marching in a single file to their community. Without the action of the ants, plants would not have been in existence today."

When Willow finished telling the story, the children felt refreshed and enlightened. Ward was enthused. The story was meaningful to him. It emphasised a danger which insect pests like aphids posed to his crops. He had no doubt that pest presence in a field crop is a bad thing which no one sensible would welcome. He learnt a good lesson from the action of the ants. He applauded the queen of the ants for her wisdom and discernment. He saw the ants as protectors of life. He likened them to himself. The only difference was that the ants had their teeth and jaws and he had his deadly poison.

Ward waited for everyone to tell a story before he gave his information. By now the younger ones had fallen asleep.

"Already you all know how far I went in search of what to do to prevent pest infestation on our crop," he spoke thoughtfully. "I travelled to various places for this but returned with disappointment. I am now happy to tell you that Floradale has the answer," he paused and his family listened expectantly. "Some time ago," he continued, "it occurred to me that in the forest are various plants whose extracts kill pests. I have checked the extracts on and on and ascertained that they work perfectly well. So, in the next planting season, I will start using them against the pests."

"Are you sure that they kill pests without affecting crops?" Willow asked in mixed relief and caution.

"I am quite sure," Ward replied. "They are specifically lethal to insects."

Although the household didn't see the pest as something to worry too much about, they were happy about the news. They felt that Ward had found something that finally erased his fear and anxiety. With feeling of relief, they went to bed.

The night darkness intensified as time passed. Repeatedly, the still atmosphere was pieced by the voice of an owl. The voice was very loud and pulsatile. Owl, the deluder was once again hooting. In the daytime, it went into hiding, causing people to believe that it was no longer in existence. Now it was hooting, booing, deriding them in their delusion. But this time, to the household, the hooting was no longer the mere derisive voice of the bird, but a sedating sound. As it pulsated, it lulled the family to sweet sleep. Then suddenly, the tranquility of the night was broken by a loud noise of chattering of fowls and barking of dogs. The household woke up and intuited the cause. Armed with spear and club, like a nocturnal animal, Ward and Edmund dashed towards the scene. Foxes sneaked into the household's poultry shelter to prey on the birds, but through the quick action of Ward and the son, the predators were prevented from succeeding.

Shortly, Willow who was carrying a burning torch, arrived at the poultry shelter. Ward was already there.

"From where did they enter?" Ward asked curiously. He took the touch and he and Willow started searching for a clue. Later on, Edmund who had been chasing the foxes arrived and joined in. Ward then saw the footprints of the foxes. The marks indicated the high speed with which the foxes escaped. He followed the trail and saw a large hole in the poultry fence. He examined the hole in detail.

"Oh! I see the reason," he announced. "Dale's expertise in fence making woke us up."

The grandson created the entrance for the foxes. He had been imitating the parents in fence making and repairing. Regularly he would go to that section of the fence, hitting it with his toy farm tool, leading to his weakening the fence and creating a small hole in it. When the foxes arrived, the hole was too small for them to enter, but they managed to get through. One of them forced its head through the hole and wriggled until it entered. One after the other, they carried out a similar technique. As they did, the hole continued to widen until it was so wide that the rest easily entered. After identifying the hole, Ward took some timbers and sealed it off. The following day, he repaired the fence.

By now, the winter season was ending and Ward was looking forward to the approaching planting season. Again and again, he would go to the forest, ascertaining the toxicity of the insecticide. The result he was getting was always promising. He felt so relieved, like a feeling of relief by a traveller who was going to The Promised Land, and who suddenly saw on the horizon a vague image and glory of the land. But was Ward's dream a true or a false one? Was the road he chose a way to the solution of his problem, or was it an illusion leading him to spark off environmental disaster of unimaginable dimension? Time would tell.

Chapter 7

Every morning, the sun rose radiantly from the east, making its way to the west. The sky's sullen face had given way to its bright blue smile. From desert to luxuriant meadow, grassland had turned, and trees dressed up in new foliage and colourful flowers as their festive regalia. The atmosphere was warm, and the air filled with floral scent. Birds had re-appeared, merrily chirruping and fluttering in the air. It was spring season. The earth had woken up in its vigour, life was full of the sound, sight and smell of nature. Revived, relieved and refreshed, all creatures were thankful.

The awaited season had come. Ward was preparing to plant his crop. It was in the middle of this that he had a visitor. After Ward had welcomed him, the visitor made his request:

> "I am Ackley, a native of Oakland. I farm with my extended family. We have a large area of farmland but we don't have enough crop to plant in it. I come to ask for your help. I beg you to lend us your grain and I promise that in the next harvest I will repay you two times the amount you give me. If you agree, I will go and come back with your in-law, Brent for him to witness my promise."

Ward thought deeply about the request. He knew that sometimes crop failure occurred. If this happened to the visitor and his family, they would struggle to repay him. Putting this into consideration, he replied: "I will help you on the condition that you will do your best to plant the crop. I am not asking you to repay me by next harvest. I will give you the crop to keep on planting until it yields enough that will sustain you and your family, and for you to pay me back just the amount I am going to give you. Go and come back with Brent for him to witness this agreement." Ackley was very happy. He thanked Ward with all his heart. He then left and soon returned with Brent.

"Who is this?" Ward asked Brent.

"He is Ackley from our community," Brent replied.

"Do you know his family and their occupation?" Ward asked.

"I know him and his extended family," replied Brent. "They are farmers and they have a large farmland. For some time they have been clearing the land. In-fact, it was a few days ago that I asked him where he kept the crop to grow on such massive land, knowing that they don't have enough in their barn. He told me that he was going to exchange his livestock for more crop."

Turning to Ackley, Ward asked: "Have you decided to cancel making the exchange?"

"I exchanged the livestock for the crop, but the crop was not enough," replied Ackley.

Without further question, Ward gave him bags of grain more than he actually needed. He then spoke to Brent directly: "You are here as a witness. I gave him the crop on the grounds that he and his family will try their best to plant it. I will only receive from him the amount I have given him after the crop has yielded enough to sustain him and his family and for paying me back."

Brent thanked Ward and then helped Ackley to transport the grain to Oakland.

The significance of the meeting was hidden from the beneficiary and the benefactor. Ward's gift to Ackley was like a loan whose repayment was due when Ward became a victim of his own design.

Shortly after the visitors had left, Ward and his workmen cleared his land, loosened the soil and began to till it with ploughs pulled by oxen. As the soil was being broken, a smell of rotting organic matter rose, filling the air. Some soil organisms that were exposed ran, seeking shelter in the crevices of stones and clods, and others flew. From the air and forest, birds espied the activity and flocked to the field, following behind the ploughs, feeding on the nutritious organisms. In the end, the farmers broadcasted their cereal crop on the tilled field, marking the end of the first stage of the planting. Well fed, the birds dispersed gratefully, singing their thanks as they resumed their various occupations.

By now, the grain had germinated, their shoots emerging from the soil. Ward waited for some time and then began gathering relevant plants, extracting the insecticide. As he was doing this, Edith and Dale arrived. One could easily notice a strange feeling in their mood.

"Grandfather what are these?" Dale asked in unease.

"They are plant extracts," Ward replied.

"What are they for?" the boy continued.

"For saving our crop," replied Ward.

"What are they saving the crop from?" Edith joined in.

"From damage by insects," Ward replied.

"Do they kill insects?" asked Edith.

"Yes," replied Ward.

"But the ants kill the insects," said Dale.

"They are not killing enough of them," Ward replied. "They leave so many of them to damage our crop."

"They don't," the children protested together.

"We always have enough crop to harvest," Edith said.

"The ants protect the crop," Dale added. "You lie, you want to kill the ants. If you kill them, their queen will send her soldiers to invade us."

Ward laughed, feeling sorry for what he saw as the children's false belief. Finding it difficult to convince them and being too busy to continue the argument, he lied to them: "Love I was pulling your leg. The extracts are for making the soil fertile."

The matter was now settled. Pacified, the children sat down and watched Ward with interest, as he extracted the insecticide.

The pesticide was now ready. With the help of his workmen, Ward distributed it to his field crop and applied it as a blanket of poison over the whole field, including his permanent pasture in which his livestock grazed. After some time, he and Edmund went to see how the poison had worked. Yes, it worked as he expected. It was like a noiseless explosion that claimed many lives. From the surfaces of leaves down to the soil were corpses of insects. None of these caught Ward's attention more than the corpses of ground beetles did. The dead beetles lay on the ground, their legs hanging in the air. The death of the beetles and the way they lay was meaningful to Ward. To him it was a confirmation of the power of the poison and a herald of an imminent end of his fear. It seemed he was already witnessing a proof of the truth of the words spoken to him. There was no doubt in Ward's mind that a glorious future was awaiting him. The hanging of the beetles' legs in the air had even a deeper meaning to Ward. He saw it as a sign of the insects' surrender to him. As he looked at the dead beetles, he nodded several times, in acknowledgement of the satisfactory action of the poison and the insects' surrender.

Nevertheless, the insects had not completely vanished from the crop. Despite the action of the poison, Ward still found some of the shoots of the crop damaged by the pests. He uprooted a few of the shoots and looked very closely at them. He then sliced them open and found fly maggots within them. The maggots were alive feeding on the shoots. He showed it to Edmund.

"Do you see?" he said in surprise. "Despite all the action of the pesticide, some of the pests still survived and caused damage. Maybe the pests arrived in large numbers and these are the ones the pesticide couldn't reach. Thanks to the second voice that spoke to me, telling me that I shouldn't wait. If I had waited, it would have been too late to save my crop."

Edmund looked closely at the maggots. "They are the lucky ones," he said.

"But this is going to be the last chance for their luck," Ward said. "In the next season, they will definitely run out of luck," he nodded in threat. "By continuously using the pesticide, applying it in higher dosage, there will be no chance of their surviving."

Ward was very confident and fully convinced that soon the worry about the pest would be over. He and Edmund went home happily. They were now looking forward to a fear-free, peaceful and glorious future, and furthermore, to a change of Ward's image to an image of a god.

Chapter 8

Ward used the insecticide on his agricultural land for a very long time. After that, Floradale had become a land of pests and soil infertility. When he began using the insecticide, he started noticing a gradual but steady rise of pest population in his field crop. Every time he noticed this, the following planting season, he increased the dosage of the insecticide. But the more he did this, the more the pest numbers increased. Like a spear which one threw at his enemy and which didn't kill the enemy, but instead provoked his aggression and reinforcement, the poisonous weapon was followed by an unbelievable increase of the pest population. But Ward persisted in the war. In response, he would again and again enter the forest and came up with broad-spectrum insecticide and applied it over the entire agricultural landscape. In this way he maintained application of the poison on his agricultural land for a very long time. Yet, it appeared as though the insecticide was accelerating the reproductive rate of the pests and enhancing their survival capacity. Ward's crops in his agricultural field were there just as a nutrient for a myriad of pests. On the stems and leaves of the crops were aphids as many as grains of sand on the seashore.

Ward's planting was like burying one's crop in a dead soil. He had lost many of his livestock and workmen.

His barn was almost empty of crop. To survive, he had to gather forest food. With fortitude, Ward faced the situation. But what he was not able to bear was the alleged cause. This worried him so much.

There was a rumour spreading within the Woodland clan. The clan suspected that Ward had done something wrong and that his suffering could be the consequence. Some hunters from the clan had encountered him in the forest several times. This time he was gathering forest food at the border of Floradale and Woodland. The hunters were shocked and so they went to their oracle, Wood, to find out what had gone wrong with Ward.

Wood was an intermediary between the clan and their deities. The deities included the god of the forest, the sky god, and the spirits of the clan's ancestors. In addition to many things the people got from the deities, they received information from them about the causes of misfortunes, diseases and deaths.

In Wood's homestead was a shrine. In the shrine were carved images and stones which bore schematic designs, all representing the deities. There were also antlers and animal skulls which were dedicated to the god of the forest. In the presence of the objects, Wood entreated the deities to make the hunters successful in their hunting, and to protect them. This was one of the reasons a proportion of the hunters' game animals were given to Wood regularly. Anyone who went to his hut would not fail to see some fresh game animals hanging on the frame of the roof of the hut.

When the hunters arrived, Fairhill, their leader spoke on their behalf. He had to speak loudly and say who they were. The sense of the old man's perception had undergone an enormous transition, from the perception of mundane things to an insight into the events and protocols of the gods.

"Our living ancestor," Fairhill addressed the oracle. "We are your children who revere you, and give you our game animal in acknowledgement of your sacred duty. I am Fairhill, the son of Wildfield. We come to gain knowledge from you. We request you to explain to us the reason behind a strange occurrence which we are witnessing these days. This is about Ward of Floradale. Everyone knew him as the most prosperous man in the whole territory.

"What is this that is happening to his crops? Pests are everywhere in his field. He has lost all his crops and livestock. Recently, we saw him almost travelling to our region, searching for forest food. Tell us what went wrong with this man. What did he do wrong?"

Wood went to the shrine and then returned in shock and with an alarming feeling. "Oh, what a bad omen!" he exclaimed in sorrowful mood, his voice quivering. "Woe to evil doers! I lifted up my eyes and saw a rainbow that lost its colour and sparkle, a faint grey rainbow hanging in a dark and cloudy sky; the environment eerily silent and cold, waiting for a storm that had never been experienced before. I enquired from our ancestors what the omen portends. They told me that a bleak future awaits us, because of what the man did. The man broke the law of nature, and this sparked off his suffering and a natural disaster that is heading to engulf us sooner or later. His action is the worst sacrilege ever committed since our forebears established the sacred seal that united the god of the forest and the spirit of our hunters."

That was it. The hunters dispersed in horror. Since then, the image of Ward in the clan was like that of an outcast. This is why Stan; a member of the clan became a symbol of ingratitude.

Ward sensed the people's belief and reaction. They deeply perturbed him. Nevertheless, he ignored them and went on with his daily life. But there was a rumour circulating throughout Oakland. One day, Ackley visited Brent.

"Did you hear what I have been hearing in our community?" Ackley asked.

"What is it?" Blake replied in unease.

"Some people are accusing Ward of profaning against the law of the fertility goddess," Ackley began. "According to them, our gods revealed this to Heartroot, our priest. They said that what is happening to Ward is the result of a curse from the goddess. Recently, I went to Meadowvalley and saw some people whispering the same story. Now I heard that the priest warned all the farmers in Oakland never to allow Ward to tread on their agricultural land, for if he is allowed, the soil will turn infertile. This is so serious that some farmers are armed with weapons, ready to kill Ward if they see him walking on their field. I come to tell you this and suggest that we go and see Ward and offer him our support, and advise him on what to do for his safety."

"Thanks a lot," Brent replied in relief. "When do you want us to go?" he asked submissively.

"We should not waste time," Ackley replied. "Let's go tomorrow, and we must take some of our agricultural produce to give to Ward." He then left and rushed to Oakland market.

When he got to the market, he went to a stall where Stan from Woodland clan was trading his agricultural produce.

"Stan," Ackley called him and spoke pleadingly. "There is a serious matter which needs our urgent attention. You are aware of Ward's situation. You know about the problem he is facing. Everyone knows how a lot of people in this land benefitted from him, and survived through him, of which you and I are an example. But in spite of this, no one bothers to show sympathy to him in his suffering. I beg you Stan; let's not behave like these people. Let's go and acknowledge his charitable work which earned us our means of living. Let's go and see him and give him some of our agricultural produce. We are able to have this because of the crop he gave us in the past. This is the time for us to thank him for our means of living which he built."

Stan was glaring at Ackley as he spoke, and immediately he finished speaking, he replied in disgust: "Man of Oakland, are you your master's agent? Has he destroyed his means of living and sent you to direct me to destroy mine? Has he sent you to come here and tempt me?" He spat in anger and went on. "Curse upon you and him if that is your aim. Let it be your abomination for uttering that what I inherited from my father is a gift from an infamous being I know nothing about." Indignantly, he turned his back on Ackley and with a sullen face looked away, unmindful of Ackley and Ward.

Stunned, Ackley gaped at him. "Stan," he called. There was no reply. "Stan!" he shouted. He was talking to a brick wall. He called on and on but there was no reply. He then got up and left.

Having returned to his home, he set apart some bags of grain and three sheep. The following day, he took the items to Brent. These were added to some bags of grain which Brent had for Ward. They took these and went to Floradale.

When they arrived, they gave the gift to Ward, and then Brent spoke to him.

"Ward," he began cautiously, "there is a serious matter which we don't want to hide from you. This is the main reason for our coming here. It is not bad news but precious information. We come to advise you not to go to Oakland for any reason. People are spreading false news that you offended the goddess, and so she cursed you, and that any soil you step on, turns infertile. They have agreed that they will not allow you to tread on their agricultural land. They are now armed, ready to kill you if they happen to see you walking on their land." He stretched out his hand and held Ward's shoulder in reassurance. "Please Ward," he went on, "to avoid trouble, stay in Floradale for the time being. Don't go to Oakland for any reason. If there is anything you feel is a cause for you to go there, send any of your children to inform us and we will deal with that. We know quite well that the belief is false. But we can't erase it immediately from the people's mind. At the moment, stay in Floradale to prevent any chance of their harming you. If you go to Oakland, although you may not go to their agricultural land, they can falsely accuse you that you did, and use this as an excuse to kill you. Please Ward, avoid Oakland at the moment. I and Ackley are working hard to convince the people that you are not what they see you as. Don't panic. We will succeed. Very soon we will come back and announce our success. But meanwhile stay in Floradale."

A tidal wave of fear overwhelmed Ward. His heart beat faster in panic. He didn't know what to say. He only stood in shock, gaping at his visitors.

"We are going," Brent said in a hurry. "Please don't worry. Take my advice. I assure you that we are ready to offer you all our support and help. I will come to see you again very soon."

"Thank you for the information and your effort," Ward replied in dispirited mood."I also thank you for the gifts". His visitors then left.

After the people had gone, Ward thought deeply about the matter. "If Oakland perceives me in this way and is ready to treat me as I heard," he said to himself sadly, "which other community will treat me better?"

Ward looked back to the past and remembered Straw's story. He came to the conclusion that the story was a true foretell of his fate in disguise; that actually, the victim in the Straw's story was him. He saw his situation as the manifestation. He saw himself as one destined to face the fate and so he resigned himself to it. Ward had now decided to take an action which even the worst outcast in Highland would not feel hopeless enough to take; a decision of which any ear that heard it in that world would tingle.

Chapter 9

For some time after receiving the message from Blake and Ackley, Ward had been having secret meetings with his wife and Edmund. There was always sadness on their faces which caused the rest of the family to suspect that something unusual was happening. In the last meeting, Ward addressed the wife and son, pointing out the only option they had.

"Following our reflection last time, it is obvious that Floradale is not our home, just as Highland is not our home," Ward said in resignation. "We are not destined for agriculture. We have to go to where our home is. We have to go to Marshland and settle there, and start hunting and gathering as our means of living. Our going there is not for a good livelihood. In-fact, we are running away from a better livelihood. Despite our hardship, we are still better off than the inhabitants of Marshland. Here we have some friends who are ready to help us in our problem. But their help can't ensure my safety in this territory. What we lack here and what we desperately need is my safety. Our going to Marshland is for ensuring the safety, although it is not guaranteed.

"What worries me now is how these children will cope with the difficulties we will be facing during the journey, and in the land. But we can't leave them behind to live in Floradale. Leaving them here is to leave them to suffer and die. If they however manage to survive, they will only be here as the inheritors of our bad name and So, they will live to suffer.

"Do you have any suggestion?" he asked, looking at Willow and Edmund for a reply.

"Lina is married and had a child in her new family," Willow replied in sad mood. "Although she is our daughter, she is now a member of another family. Her child is a child of Oakland and so she is a member of Oakland community. So, she can stay with her parents-in-law. Tiller and his Moorland fiancée are not yet married. He can leave the poor girl and go with us. The only ones I worry about are Edmund's children…"

"And his wife, Mildred," Ward interrupted. "Be aware that we are going to face difficulties in the journey and in the land. But we have no choice. Mildred has a better choice than us. She has her people who are ready to live with her. She is a daughter of Oakland and so the people won't see her as people around see us. To save her from the suffering which we are going to go through, we must let her go back to her parents. We can take her son and daughter and pray that they survive with us."

It was a difficult suggestion to Edmund. Although he was in full support of the whole plan, he hesitated in accepting the suggestion. Separating him from his wife would be more painful than his suffering together with her and the rest of the family. However, he understood the level of suffering involved in the journey. He was not willing to advise anyone to go with them, when one had an alternative and a better place to go to.

"I am fully aware of the level of suffering we are going to go through," Edmund responded in distraught to the father's suggestion. "I know very well that Mildred will be safer if she stays with her people. But I am also aware of the distress which our separation will cause her and me. I can't agree to this suggestion at the moment. I need to talk to her and find out her choice."

They all agreed but before the end of the meeting, Ward and Willow pleaded with him to persuade his wife to return to her parents.

For days, Edmund tried to persuade his wife not to join them in the journey. But each time, he failed. In-fact, his wife was more determined than any of them to make the journey. Her bond with Edmund was like the bond between Willow and Ward. She couldn't live happily without Edmund and their children.

Edmund had brought his finding to the parents. Ward then gathered together the whole family in his hut.

"I call all of you to inform you about a danger that is around us and how to save ourselves from it," he addressed the family. "The soil of Floradale has turned against us. Today, our happiness is replaced by distress. People of Oakland have developed a bad belief about us. They have concluded that our crop failure and our sufferings are a result of a curse placed upon us by their deities. Today, we are no more seen as just people of hard life, but also as desecrators. They are now armed, ready to kill me if I go to them.

"We have suffered enough in this land, and so we have decided to leave Floradale for a land far away."

The children's mood became despondent. There were tears in their eyes as they were apprehensive of losing their friends.

"The journey will start after this winter," Ward continued. "I and Edmund will exchange the few livestock we have for foodstuff and other essential items we will need in the journey, and for settling in the land. So, this winter will be our last winter in Floradale." He then turned to Mildred. "I beg you my daughter," he said to her. "Save yourself from the suffering we are obliged to go through. Please don't join us. Stay with your people, and live in peace with your parents. We will take your children with us. We will look after them. Stay and maintain good relationship with Lina. Try to comfort her in the sorrow which our absence will cause her."

"If you stay behind, it will be of more help than going with us," Willow added pleadingly, sobbing.

Mildred had been listening in an unmoved mood. "I've heard your advice just as I heard Edmund's", she replied. "The most distressful thing to me is losing you, and this is the worst illness which I can't bear. Wherever you go for any reason, there will I go. Whatever you leave behind is what I leave behind. I am ready to go to the ends of the earth to keep you company, and there, your sorrow shall I share with you, and your joy shall be my joy. Where you die, there will I die. Your grave shall be mine and where your spirit goes, there shall mine go."

The die was cast; the meeting was over and the journey was inevitable. They dispersed with feeling of anxiety, waiting for the end of the winter.

But one day in the afternoon, a dull red cloud filled the sky as a night owl shrieked. The atmosphere was eerily still and chilly.

"These must be omens of doom," Ward muttered in fear.

Then suddenly, the sky darkened, followed by a flash of lightning and a simultaneous crash of thunder. The bellow was so powerful and loud that it threw Ward down before fading, like the fading of the sound of a vehicle as it travels away, its pitch steadily lowering. The faint rumbling sound was now as sedating as a quiet deep sound of musical rhythm – the type of sound that transposes one's mind into states he might usually have difficulty reaching, allowing him to experience what those states feel like. It was at this stage that Ward was in a deep trance. In the trance, he heard a gentle, clear mystical voice.

"Ward! Ward!" the voice called. "Behold, it is written: the animals supply the needs of man; the plants supply the needs of the animals; for all the soil organisms, bits and pieces assembled together, mimic the ecosystem in order to supply the needs of the plants.

"Will man exist if he kills the animals? No, because he has removed the supplier of his needs. Will he be sustained if he kills the plants but saves the animals? No, because the animals will die, following the death of the plants. Will his needs be met if he saves the animals and the plants but destroys the soil organisms? No, because he has destroyed the root of the whole system. This is what you have been doing. You have been destroying the soil organisms, the taproot of the whole system.

"Look, Floradale is like the system. It stands as a symbol of nature in its perfect order. Like a living tree, it stands. As a taproot sustains a tree, so do the soil organisms sustain the land. As a tree trunk and the branches are part of a tree, so are the plants part of the land. As a tree bears its fruits and seeds, so does the land bears all that sustain you, including your crops and livestock. As a bird sits on top of a tree for shelter and nourishment, so do you sit on the yield of the land.

"You were equipped and placed in this position for a purpose. You are there as a custodian, a manager, a caretaker whose role is to work in harmony with all the organisms that are part of the system. By carrying out the duty, you will reveal to all people the position and duty of man in nature – that man is at nature's pinnacle, where he is designed to be, to sustain and be sustained.

"Return to your soil and stop the application of the poison on it. Watch the soil and see how the organisms, the taproot of the life of the land recovers as the wound heals. Study the organisms and their interactions, and learn from your mistake."

When Ward woke up from the trance, the memory of the effect of the poison on the soil insects, especially the ground beetles came to his mind.

"Could they be the ones that are the root of the life of the land?" he mumbled in suspicion. He remembered how their dead bodies lay, raising their legs. "Could it not be a sign of declaration of their innocence to the creator rather than a sign of surrender? What are these insects and what is their role?"

Ward had asked a question, of which his search for the answer was the beginning of his journey to his knowledge of interrelationships of organisms and the significance; a knowledge that revealed the root cause of his problem.

After ruminating on the revelation, he went to see his family. He met Willow at the domestic area. He was cheerful and relaxed.

"What changed your mood?" Willow asked in surprise. "Is it the strange weather today?"

"I have good news for you," he told the wife with a lively voice.

"What is the news?" she asked expectantly. "Please tell me. Have Oakland people changed their thought about you, and cancelled their resolution?"

"I didn't go to Oakland, neither did I see anyone from there," he replied and then gave his message.

"Tell Edmund and the rest that I will like to see all of you after supper for the details of the news." He then returned to his hut.

The family had finished their supper and were in Ward's hut. In despondent mood, they waited for Ward to speak. He had meditated on the matter before he began.

"We planned to leave Floradale after the winter, for settling in Marshland," he said. "I call you to inform that there is a need for it to be suspended. I received a sign that indicates that we might finally abandon the plan."

The family heaved a sigh of relief. Edmund shook his head in disbelief and asked: "Have Oakland people changed their mind and if so, from whom did you hear that?"

"No," replied Ward. "The information I have for you is not from them."

"Tell us the sign you received," Willow urged impatiently.

"I was in this room today," he went on. "From here I travelled to a zone far beyond human reach and received a good message. It was like the message I received in the forest of Highland. It revealed to me a step I need to take for the solution of our problem."

"Please tell us the step?" Edmund requested.

"Like the instruction that led to my settling in Floradale," said Ward, "the message instructed me to embark on a project which requires patience. There are certain organisms in our soil. They are said to be the root of our livelihood, and that the pesticide I have been using on the soil was killing the organisms. I was told to stop using the pesticide, and that I should study the interactions of the organisms; by doing this that I will develop an insight into the solution of our problem. I will wait for some time before starting the project.

"Now, this is the time when the charitable work of Ackley and Brent is most needed. As we don't have a crop to start planting in the soil if the problem is solved, the ones Ackley and Brent will be giving us will enable us to do the planting." Ward was talking about the plan which Ackley and Brent had for him.

After the time they went to warn Ward not to go to Oakland, and comforted him with some gift, Ackley told Brent that he and his household had reserved a piece of their land for growing their crop for Ward. Brent was very pleased and he promised that he and his family would support in the planting, and in harvesting and transporting the crop to Ward. They informed Ward about the plan. Ward thanked them, but in his mind, he didn't feel that the plan was the right thing. He felt that it was better for him and his family to settle in Marshland than living on people's charity and in fear. Now that he had the hope of continuing his agricultural occupation which could lead to a disproof of the people's belief about him, he saw the crop as very important.

Ward's information had given the family a hope which displaced their anxiety. They went to bed with a feeling of relief, praying that the dream came true.

Chapter 10

Following the revelation which Ward received, he suspended planting for a long time. By now his agricultural field was like a forest of wild plants. Then he cleared it, tilled the soil and planted his crop in it. Part of the crop was the one Ackley and Brent were supplying him. The planting was just an experimental work for studying the organisms in the field. He was not using the pesticide anymore. Initially he noticed that the pests were still there in large numbers, but at the same time, other organisms were emerging, including the ground beetles. As time passed, he began noticing that as the other organisms increased in number, the pest population decreased. This reached to a stage where the pest population was no longer falling nor rising, and the population of the other organisms was no longer rising and was not falling. It was at this stage that he noticed that the damage of the crop by the pests was not significant. He was now able to plant his crop and reap a good yield.

One day, Ward took a wooden bowl to his field. He picked a few of the ground beetles and put them in the bowl. He then uprooted few shoots of his crop in which pest pupae and maggots were. He sliced the shoots and collected the pupae and maggots. Having them in his palm, and showing them to the beetles, he mumbled to the beetles: "What is your role in my soil? These pupae and maggots have told me what they do. They are the eaters of my crop. They are my enemy. Are you here as a friend or as an enemy? Just tell me." As he questioned, some of the pupae and maggots suddenly fell into the bowl, and instantly the ground beetles started eating them." The secret was revealed!

"Oh!" Ward exclaimed in deep surprise, "the eater of the eater of my crop. So, all the time it was my ally that I was killing, thinking that I was killing my enemy."

It didn't take the beetles much time to consume the meal. Ward then added the rest of the pupae and maggots in the bowl for the beetles to eat before releasing them back to the field.

Ward's insight into the habits of the rest of the organisms in his agricultural field was deepening. He noticed aphids sucking the sap of the leaves and stems of the crop. He observed ladybirds feeding on the aphids.

The root cause of Ward's agricultural disaster was gradually becoming clear. Insect pests feed on crops, but they have their natural enemies that prevent their population from rising to a level that would lead to the destruction of the crops. The natural enemies include the ground beetles and ladybirds which feed on the pests. They also include some parasitic wasps that lay their eggs in the pests, using the pests as a nutrient for rearing their offspring.

In every natural environment, there are different groups of organisms, each playing a particularly important role. The population of each of the groups is maintained at a specific level for the group to play the required role, and not to overplay or underplay the role, because overplaying or underplaying the role is not favourable to the environment. It is the natural enemies that maintain the populations of the organisms at the required levels. The population of the insect pests is one of the populations which the natural enemies maintain at the required level: If the population of any of the organisms rises above the required level, the natural enemies intensify their predatory and parasitic activities, leading to a fall of the population to the required level. On the other hand, if the population falls below the required level, the natural enemies relax their activities, leading the population to rise back to the required level.

The population of the natural enemies is also maintained at a required level by various factors; one of which is scarcity of food: If the population rises above normal, there won't be enough food to maintain the natural enemies, leading to their population falling back to the normal level.

The cause of Ward's problem was that the pesticide he was using was killing the pests' natural enemies more than it killed the pests. Also, the ability of the pests to recover from the poison was greater than that of their natural enemies. Usually, any reduction in the number of their natural enemies, even if the pests too were reduced to the same proportion extent, puts the pests at an advantage.

So, all the time Ward was using the pesticide, the poison was acting in favour of the pests. It was decimating the pests' natural enemies, making it possible for the pest population to grow out of control, leading to the agricultural disaster.

Having developed insight into the importance of the natural enemies, Ward resolved to do everything he could to protect them. He then turned his attention to plant litter on his soil. He noticed a good number of small organisms on the litter. The organisms vanished during the time he was applying the insecticide on the soil. The organisms included oribatid mites. He noticed that the plant litter was rotting very fast. It was not like that during the time he was using the insecticide. During that time, the litter hardly rotted. It was only there like a preserved material. As an experienced farmer, Ward knew that rotting of plant litter was usually followed by soil fertility. He concluded that the organisms were important in the rotting of the litter and therefore should not be harmed.

Yes, Ward was right in his conclusion. The organisms are decomposers. They feed on organic matter, including plant litter. So, plants give them the litter to feed on and through the feeding, they decompose the litter, releasing plant nutrients in the litter, making the nutrients available for the plants to absorb with their roots. The oribatid mites play important role in the decomposition. But when Ward was using the insecticide, the poison decimated the organisms, thereby preventing the decomposition of organic matter in his soil, leading to the soil infertility and therefore stunted growth and poor yield of his crop.

Ward then turned his attention to the bees and butterflies on flowers in his field. During the time he was using the insecticide, these insects disappeared, but now had re-appeared. Experience showed him that after interaction of insects, like the butterflies and bees with plants' flowers, what followed was the plants bearing fruits and seeds. He felt that the vanishing of the insects during the time he used the insecticide was because the poison killed the insects, leading to the poor yield of his other crops that needed the insects for their seed and fruit bearing. He affirmed that the insects played a beneficial role and must be protected.

Yes, Ward was right in his affirmation: the other plants and the insects depend on each other. The plants' flowers give the insects nectar and pollen as food, and in return, the insects pollinate the flowers, making it possible for the plants to bear fruits and seeds.

Putting these together, the hidden image of Ward's agricultural field became obvious. It was now clear to him, that his field was a natural environment where organisms interacted with one another for survival; a system of interdependence of living things where plants produced food for man and animal creatures, and in return, the creatures functioned to maintain conditions for the plants' well-being and existence. Indeed, it was a system where the functions of animal creatures ensured the supply of the needs of plants, making it possible for the plants to supply the needs of animals, and the animals to supply the needs of man.

Following this insight, Ward accepted the pests as the natural part of organisms in his agricultural field and that one should not worry about it, except when the population rose above usual in the environment. He agreed that in controlling the population, he must use an appropriate insecticide whose action would only complement the action of the natural enemies: he would adopt a selective insecticide and apply it at such a time of the season and in such a way and dosage that it only acts as an agent auxiliary to the function of the natural enemies.

The finding had sparked off Ward's curiosity about the real nature of the forest. He started reading meaning into the interactions of diverse organisms in the forest, one of which was the interaction of birds and plants. He perceived mutual benefits in the interactions and concluded that all organisms depend on each other for their survival in the forest just as it was the case in his field crop.

Ward's conclusion is in agreement with the interrelationships of diverse organisms in the forests in different parts of the world. The forest plants include those that grow on the surface of trees, as well as those that climb on other plants. Although these plants seem to be weak, their existence in the forest is indispensable to the life of the forest. They contribute significantly to the survival of the forest animals and plants. The plants that harmlessly grow on the surface of the forest trees, offer food and shelter to small animals, including insects. Without the plants, the forest wouldn't be able to support the diverse amount of animal life in the forest. The plants that climb on trees produce abundant fruits, leaves and flowers for the forest animals to feed on. In addition, by climbing the trees, they form networks of bridges that are linked to different trees in the forest. Through the bridges, some animals spread plant seeds evenly in the forest for successful germination and growth. Also, through the bridges, animals move from tree to tree to get their food. Without the bridges, the animals would have to climb down from one tree before climbing up another for getting their food. This would then expose them more to predators on the floor of the forest, leading to the predators reducing the population of the animals below the required level in the forest.

As a forest gets older, plants in it compete for space, light, water and nutrients in the soil. This would lead to the forest extinction. But this problem is averted by the mutualistic relationships that exist between the forest plants and some animals like birds. The plants call the birds and give them fruits, saying: "Take these as your food and pay us back by taking the seeds in them to a suitable environment for their successful germination and flourish." The birds sing their thanks and then gulp down the fruits without chewing the seeds, digest the freshly pulped fruits and defecate the seeds in a suitable environment, away from the forest where the seeds germinate and flourish as a new forest. This interdependence of forest plants and animals demonstrates that the image of the forest is similar to that of agricultural field, or that the forest is a community of organisms where the organisms interact with each other and with non-living things which include water and air that surround them for their survival. It is this system that is known as the ecological system or ecosystem – a system where the life of one organism depends on the life of the other and the non-living things that surround them.

Being aware that the travel of birds to the forest to feed on the fruits in the forest, benefits the birds just as it benefits the forest, Ward guessed that a similar mutual benefit could arise when seabirds travel to the sea to feed on the fish in the sea.

Ward was right in his guess. Just as land has plants that are food for the land animals, the sea has microscopic plants on the surface of the sea which are food for small organisms in the sea. The small organisms in turn are food for bigger organisms including the fish which the seabirds feed on. But the sea needs the land to provide nutrients for the sea plants, and seabirds which nest on the land need fish in the sea as their food. So, the sea offers the fish to the seabirds as a medium of exchange, just as the forest offers fruits to birds. The seabirds then take and eat the fish, and defecate their droppings on the land. The droppings of the seabirds contain very important plant nutrients. This is because of the fish which the seabirds eat. When the rain falls, it washes the nutrients into the sea for maintenance of the sea plants. This is a demonstration of interdependence of life in the sea and life on the land, even thoug the land and the sea are separated from each other.

With this experience and insight, Ward came to the conclusion that no life exists in isolation, independent of the others, and that any harm of one organism is a harm of the rest of the organisms. He saw what he did on his field crop as a crime against all life, including himself.

He suspected that the frightening weather condition; the shrieking of the owl; the sun's sudden withdrawal of its light, and the thunderbolt which he experienced, were an expression of the anger of the earthly and heavenly forces against him.

After considering the role of the thunder lightning and the sun in maintaining life, one can agree that Ward was right in his suspicion. The thunder lightning supports the life on the earth by striking nitrogen gas in the air, turning it into a form that finally becomes a plant nutrient in the soil. The sun gives light which plants use in producing oxygen and the food that gives energy. With the light, water and a gas called carbon dioxide, the plants produce the food and oxygen for their use and animals' use. When animals including man take in the food and oxygen, they gain the energy and breathe out carbon dioxide as a waste product. But the plants need the carbon dioxide for producing the food and oxygen, and so they take it from the animals. Therefore, animals depend on plants for their food and oxygen, and in return plants depend on the animals for carbon dioxide. But this interdependence is only possible because of the contributions from the thunder lightning and the sun. So, it is reasonable to agree with Ward in his suspicion. It looked like the thunder and the sun were upset because Ward was destroying their work.

Following this insight, Ward became aware of why every individual is different from each other, and why they always interrelate. They differ from each other just as a food producer and a water supplier differ from each other. Without food, water supplier won't exist, and without water, food producer won't exist. They interrelate so that their difference from each other can be meaningful. When the food producer and water supplier come together, the food producer gains the water he lacks, and in return the water supplier gains the food he lacks. This is a demonstration of the importance of diversity and uniqueness of characteristics in the interdependence of organisms.

With this knowledge and the application, Ward regained his glorious status. In addition to supplying crops to Ward which enabled him to re-establish his agricultural career, Ackley and Brent recruited employees for him. The employees took part in planting his crops and in trading his agricultural produce in communities. One of the ways Ward showed his appreciation of what Brent and Ackley did for him was his training Ackley's son, Hill, and Brent's son, Edward, in herbal medicine. For a long time now, the young men had been doing well in healing the sick in Oakland. It was this that was going to pave the way for Ward's reunion with the People of Oakland.

In addition to revealing the cause of Ward's problem and the solution, his experience contributed to a revelation of the rationality of the religious belief of the ancients and the concept they held from time immemorial:

> Right from the beginning, the ancient people believed that the nature is a created order and so, they treated it with reverence. They believed that all creations are the symbols of the eternal power and divine nature of God. They held the concept that all life is interdependent, that nothing exists

in isolation and so, the real nature of individuals and events can only be correctly understood in the contest of their connections with all others. This belief is held by Buddhists. The belief is also expressed in the Gaia Theory. Gaia was the ancient Greek goddess of the Earth. The ancient people held the concept that the interdependence and the functions of all creation on the earth and in the sky are a response to a divine order. For this, the people developed the cult of earth goddess, and cult of the sky god as part of the system through which they venerated the creations as deities. Some evidence of this includes the ancient Greek cult of earth goddess; the ancient Egyptian cult of the sun god.

Information from our today's religious book, the Bible has helped in revealing the rational reason behind the belief and behaviour of the ancient people:

- Sirach – Ecclessiasticus – Jerusalem bible, chapter 42, verses 21-25: "He has imposed an order on the magnificent works of his wisdom, he is from everlasting to everlasting, nothing can be added to him, nothing taken away, he needs no one's advice. How desirable are all his works. How dazzling to the eye! They all live and last forever, whatever the circumstances all obey him. All things go in pairs, by opposites, and he has made nothing defective; the one consolidates the excellence of the other; who could ever be sated with gazing at his glory?"
- Sirach – Ecclessiasticus – Jerusalem bible, chapter 43 verse 2: "The sun, as he emerges, proclaims at

his rising, 'A thing of the wonder is the work of the Most High!'
- Roman Chapter 1 verse 20 NTL version; "For ever since the world was created, people have seen the earth and sky. Through everything God made, they can clearly see his invisible qualities – his eternal power and divine nature. So, they have no excuse for not knowing God."
- Psalm chapter 19, verse 1; "The heavens declare the glory of God; the skies proclaim the work of his hands."
- Psalm chapter 84 verse 11; "For the Lord God is a sun and shield; the Lord bestows favor and honor; no good thing does he withhold from those whose walk is blameless."

The biblical information is in agreement with the belief that the ancient Stonehenge in England was built for religious purposes. The information seems to suggest that the Stonehenge and its connection with the solstice is a divine system that came into existence as a result of the ancient people's divine insight, and that through the system, the people interrelated with God: In the right season, the sun, one of the symbols of God's eternal power and divine nature, attains a position in the sky from where it sends its rays like a heavenly signal to the sacred Stonehenge. The people flocked to the Stonehenge to watch in awe, worship, gain spiritual fulfillment and celebrated.

The biblical information also seems to give the reason behind the Newgrange in Ireland and the alignment of its passage and chamber with the rising sun on the mornings around the winter solstice.

Ward, the great ancient naturalist, his knowledge and the biblical information have combined to unravel a great mystery. The biblical information and Ward's knowledge of interrelationships of organisms and the significance have complemented each other and unveiled the rationality of the culture, concept and religious belief of the ancient people, thereby showing that a puzzle which can't be totally solved by only science or religion can be totally solved when science and religion complement each other. This is an additional proof of the validity of the saying by Albert Einstein: "Science without religion is lame, religion without science is blind."

Chapter 11

It was a worrying situation as the health of Heartroot, the Oakland priest deteriorated. Oakland healers tried their best to heal him, but failed. Owldown healers who were called also tried but failed. In despair, Oakland men sat beside him, thinking about what to do next.

"This is a hopeless situation!" Hatt lamented. "If after all the efforts of the healers, his condition remains as it is, and no one has any idea of what to do next, are you sure that sooner or later, we are not going to lose him?"

"We don't pray for that," Brereton replied dismissively. "What men can't do; the gods can do. What is beyond our perception is very visible and close to them. Let's not relent in our prayer. Eventually they will hear us and step in and heal him by themselves, or reveal to us one who has the power to heal him."

Brereton's words reminded Hatt something. "Yes, yes!" Hatt interjected with a feeling of hope. "I agree with you. Sometimes what is hidden from the elders, the gods reveal it to a child. It has now come to my mind, that there is a lad in our community who has been healing the sick with herbal medicine in this community and our neighbouring communities. I heard that he is

Ackley's son. Could he be the one through whom our gods will answer our prayer? Let's find out."

They sent a messenger, Albert, to ask Ackley to send his son, Hill to them for a talk. Very soon, the messenger returned with Hill.

"We heard that our gods blessed you with healing power," Hatt addressed Hill, "to the extent that you have been healing people in and outside our community. Is that true?"

"Yes," Hill replied shyly.

"Did any of the people you healed have the type of illness which Heartroot is suffering from?" Hatt asked. "He vomits, and intermittently defecates watery stool. He has a high body temperature."

"Yes," replied Hill. "But the people are from our neighbouring communities. The problem is that the type of medicinal plants for the healing are very rare. They are not in our forest. I need to search for them in a land far away".

There was a feeling of relief in the men's mood. They prayed to their gods, blessed Hill and begged him to go to the land as early as he could. Hill left and the following day he travelled to Floradale.

Hill was crafty. Actually, he hadn't come across any patient with that type of illness before. He went to Floradale to tell Ward about his meeting with the men and to seek his advice. If Ward wanted him to heal

Heartroot, he would give him the appropriate herbal medicine. If, however he didn't want him to heal the man, then Hill would return to Oakland and tell the men that he couldn't find the medicinal plants.

When Hill arrived, he narrated to Ward why he came. Ward thanked him for not telling the men that he was going to Floradale for the medicinal plants. Then he asked him the details of Heartroot's health condition. Hill told him. Ward then went to the forest and soon returned with some medicinal plants and gave them to Hill. He told him how to prepare and administer them. He also gave him some fruits and mixed juice which worked with the medicine. He told Hill not to let anyone know that he got the medicine from Floradale, neither should he let anyone know that he was the one who gave him the medicine.

Hill took the items and returned to Oakland and began healing Heartroot. Since the treatment, the patient's health had been improving steadily. Regularly, Hill went to Floradale and returned with more herbal medicine to continue the treatment. Finally, the weak and frail patient was now healthy and sound. The people of Oakland were astonished and joyful. Their joy was not just because their priest was healed, but also that they had a healer in the community who had brought honour to them by showing that what Owldown lacked, was in their community. For this, the elders and prominent men in the community set up a date to visit Ackley's family to show their appreciation.

When the day came, they went with bags of grain, new agricultural tools and some livestock. When they arrived, Whitford, their spokesman addressed Ackley on their behalf:

> "You might be surprised seeing us in your place with these items," said Whitford. "Sometimes a surprising occurrence is followed by a surprising response. Your son has surprised us. His deeds have proven that he is a god-sent who saves lives and makes us joyful and proud. He has proved that what Owldown can't afford, can be affordable by us. To cut it short, look at Heartroot sitting here, he is in good health. Without your son, this wouldn't be possible, instead he would be lying in a grave and we would be mourning. These are a gift we have for you in acknowledgement of what your family did for us." He showed him the items they brought.

In a brief silence, Ackley ruminated on the matter and then replied in calm mood: "Whitford, your voice is as sweet as that of the nightingale. It is as soothing as the gentle breeze of the summer season. Oakland and sons, I and my family welcome you and acknowledge your thanks and gratitude. But I must be sincere; you are in a wrong place. You are giving your present to a wrong person.

"How did Hill become the healer you call him?" he paused, looking around for an answer. There was no reply. He went on: "What you call Ackley's household, who built its means of living?" He looked around again before continuing. "Hill is still a young man. He has not yet attained the age and status of receiving sacred gifts from our ancestors. Our gods have not yet given him healing power, neither have they bestowed upon him an insight into the disease process. Yet everyone agrees that Hill has a great healing power. From where did he get the power? I will answer the question. He got it from the living deity whom you chased away. The healing power which Hill has came from Ward of Floradale," he paused and the audience looked at each other in surprise.

"It is the healing knowledge which Hill received from Ward that made it possible for him to heal Heartroot. While Ward was ousted by Oakland," he was almost shouting his words, his voice sharp and harsh, "in his loneliness, he suffered, training a son of Oakland to be the community saviour. During that time, there were some people outside this community who maintained their good relationship with Ward. But Ward preferred your son to the sons of those people as one to be trained. He trained your son as a healer without any charge. Ward left his medicinal plants for Hill to use in healing Heartroot. It is the fruit of Ward's gracious deeds that we are enjoying, rejoicing about. Which of you can then disagree that Ward is a blessing to us? Who in his right sense will not acknowledge him as the most precious gift any god can give to a society?

"If we look back, we can see clearly the status of my household. Was the household not one of the poorest in our community? Was our career not a mixture of agriculture and gathering of forest food? You shouldn't forget that we never had enough crops to plant in our soil. But today we are known as wealthy farmers. Did we attain the status through the help of any of you in the community? The answer is no. It was Ward's charitable deeds that enabled us to attain the status. In our suffering, I sought help from Ward. Like a beggar, I went to him and laid my problem before him. Out of compassion, he gave me a large amount of grain and with this we established ourselves as you see us today."

As Ackley spoke, the men bent their faces down in shame and regret. "My kindred," he continued in a friendly mood, his voice now sounding like that of an adviser, "you are in a wrong place. If you are truly happy about the healing of Heartroot and want to express your thanks and happiness, you should go to Ward. It was his charitable work that led to the healing of Heartroot. I and my family rejoice with you for the healing, but we are not going to take the present. You should give it to Ward. I am ready to add more present to it, and carry them with you to Ward."

Killingworth, one of the elders addressed the people. "Kindred," he spoke earnestly. "Ackley said the truth. No one here, I am sure will disagree that the deeds of Ward saved and enriched people far and wide. We ousted him not because he wronged any of us. It was because we felt he wronged our gods and that the consequence upon him might affect us. But nowadays, we are witnessing things that indicate that he is an agent of our gods. Are we then right in our belief and behaviour?" he looked around and everyone was nodding in agreement with him. "I warn that we must be careful in the way we see events around us. We must not let our fanaticism mislead us. Let's follow the will of our deities, and not our own will and assumption.

"It is now obvious that Ward is a man through whom our gods help us. The recent event has proved this. We must not waste time in responding to Ackley's suggestion. Let's go and offer our thanks to Ward and apologise for our mistreatment of him."

The men applauded.

Heartroot got up and spoke. "Men of Oakland, I greet you," he began submissively. "I strongly agree with Killingworth. We should not be the leaders of our gods. We should be their followers. I accept that we were misled by my limited insight. The result was our mistreatment of an innocent and just man. Our gods required us to treat Ward as a precious gift from them, but we did the opposite. With all my heart, I beg all of you for us to go and reconcile with Ward. I thank Ackley for his suggestion and I request him to lead us to Floradale." He turned to Ackley and saw him nodding in agreement, saying that he was happy to take the lead.

Finally, they agreed on a date to see Ward. After the people had gone, Ackley went to Brent who was not in the meeting. He told him the arrangement and asked him to go with them. Brent thanked Ackley and accepted the request. Then they went to Floradale and told Ward about the planned visit and the day of the visit. Ward told his family about this and they agreed to entertain the visitors very well.

A day before the visit, Ward and Edmund dressed up and went hunting. They wanted game meat as part of the food for the visitors. It was a good day for hunting.

The day was warm and bright. With their hunting weapons, they headed towards Floradale forest. As they entered the forest, the smell of domestic life and farmstead gave way to the forest fragrance. Here and there, wild flowers and ripened fruits hung on display. Colourful butterflies flew around. The buzzing of bees and chirrups of birds mixed together, creating a soothing sound. The hunters' faces beamed with a smile of delight. They had arrived at an important spot – a piece of luxuriant grassland. There was a lot of information there. Instantly they stopped and looked closely at it, silently reading the information in details. The grass was freshly grazed, and the smell was there. There were animal droppings and the footprints of the animals.

"They are red deer," Ward whispered, "really big ones."

"They are not far from here," Edmund said cautiously. "They must be lying down somewhere very close, chewing their cud."

"This is their trail," Ward identified. "It leads to the woods ahead," he said, pointing towards the location of the woods.

Silently, swiftly, they followed the trail and entered the woods. Indeed, the game animals were at the suspected site. The hunters killed and took the ones they wanted.

By the time they were returning home, the sun had gone down and the forest vibrant life returned to quietness and serenity. All the time, a gentle wind was blowing upon the sweating hunters. In the peaceful atmosphere was a soothing song of a bird. But it was difficult to discern where it was coming from. They looked at trees around, searching for the bird, but couldn't see it.

"It is the voice of a nightingale, but where is the bird?" Edmund said, wondering.

"I don't know," Ward replied as he continued looking around.

Repeatedly, in unison, the voice of the bird and the wind rose and faded. Finally, Ward identified the site of the elusive voice.

"Yes, it is coming from this place," he suddenly announced humorously, pointing towards a bushy tree which was a stone's throw from their position. "This is where it hides. As usual, it is hiding, so that other birds won't enjoy its song." Both laughed as they remembered the story of the bird and the origin of its sweet song.

The good atmosphere, the sweet song of the bird, and the refreshing memory of the origin of the song, made the hunting enjoyable. But unknown to Ward and Edmund, they were actually a good omen that signalled a reunification between Ward and Oakland people, and imminent Ward's completion of his mission on earth.

The following day, Ward's household was in a festive mood. Lina and her family, including her husband and his parents were invited and they came early. Mildred's parents were invited and they arrived. Then, afar off, they saw men with their horses and carts approaching. There was no doubt that they were men of Oakland. Ward, Willow and Edmund went and received them, and brought them to an open space in their homestead. There were a lot of seats there. The men sat down. Ward then called some of his workmen and Willow and took them to one corner of the homestead where they whispered together. What followed the whisper were Willow, Lina and Mildred carrying giant cauldrons from their storage to the kitchen area, some of the workmen leading some livestock to the slaughter house, and some carrying a quite good number of old vessels to a hut near the open space where the visitors were. The visitors knew quite well that the vessels contained good beer. They knew because beer brewed in old vessels was better than the one brewed in new vessels. This was because old vessels had cracks and crevices in them where micro-organisms settled for long as their home. The micro-organisms were good in fermenting grains, turning them into beer.

There was no doubt in the mind of the visitors that a feast was imminent. Their eyes sparkled with anticipation. One could easily notice that they were very happy. They were conversing humorously, laughing. Parker was in the full mirthful mood. When he saw Willow passing through, he jovially exclaimed: "Ah! Willow, is that you? When we came, I saw others but didn't see you. I was wondering where you were. You have changed a lot. You now look like a queen of a great land far away."

The visitors laughed vivaciously, saying in reply: "Not like a queen of faraway land, but the real queen of our land."

But Parker forgot that when they came, that Willow was one of those that welcomed them, and that he greeted her, saying that they missed her and the family in Oakland, and that Willow replied to him, saying that they also missed them in Floradale. Willow knew the reason behind the men's happiness. She didn't want to embarrass Parker by correcting him, and being too busy for any conversation, she politely replied, "Thank you, I feel flattered."

Shortly, Ackley and Brent went and called Ward to hear from the visitors why they came.

When Ward came, Ackley spoke. "These are the people of Oakland," he said. "They include our priest and elders. They came to acknowledge you as their saviour. This is our priest," he pointed at Heartroot. "He is alive today as a result of the training which my son received from you, and the herbal medicine you supplied him."

"Are you the one whom Hill was telling me about his illness?" Ward asked Heartroot in a friendly tone, pretending not to know.

"Yes," Heartroot replied submissively. "I am the one whose life was saved by your generosity, and the one who made you a victim of his limited knowledge and insight. We are here to thank you and to apologise for our mistreatment of you. My illness was the worst ever known in our community. The cure was beyond the skills of all the well-known healers. The Owdown healers failed to do anything about it. Things reached to a stage where I had to resign myself to my fate. To Oakland astonishment, the son of Ackley came and healed me. We wondered how he acquired such skill. His father told us that you trained him to be what he is. He also told us that the medicinal plants and other ingredients which he used in the healing were given to him by you. We also heard that you came in direct contact with the gods and received from them great knowledge and insight with which you enlighten people and solve their puzzles.

"Ward," his voice shook with regret and pleading. "We have been too blind to perceive what you are to us and our neighbouring communities. Despite your improvement of our economy, and the help you offered to the poor, we still remain blind to your importance to us. This weakness in perception led us to carry out an unjust action against you and therefore against ourselves. We wrongly saw you as one who offended our gods and thought that your problem was the consequence. As a result, we broke our relationship with you. I bear the full blame for this injustice because I was the leader in the misconception.

"We are now fully aware of our folly. We have realised that you are an indispensable gift from our gods and the one whose conduct pleases them. It is clear that what we have been doing against you was a blockage of the channel through which help from our gods comes to us. We treated you wickedly, acting against ourselves.

You had your problem, and it was our duty to help you, but we rejected you. Then we had our own problem, it was a time for you to get revenge, but you came and rescued us like a merciful god. We the sons of Oakland beg for your forgiveness. We implore you to accept our apology and receive our thanks." He turned to his group, expecting their support.

"Yes," Hatt said, addressing Ward, "Heartroot said the truth. What you did for us can't be overemphasised. You brought life, wealth and honour to our people. We apologise for our failure to appreciate this, and for our foolish action against you. Forgive us our mistake and reconcile with us. We brought these for you," He pointed at the presents they came with. "They are nothing when compared to what you have done for us. Accept them from us, not as a repayment but as a sign of our gratitude for your benevolence."

In brief silence, Ward thought deeply about the people's request, and then nodded in satisfaction. "Oh, my goodness!" he uttered, lifting up his face. "Wisdom and insight have descended upon the sons of men, and they are now acting accordingly." He then turned to the men and addressed them in deep gratification. "As you came here, so have I been wishing to go to you. But I didn't because I feared for my life. I would have come to tell you where we all got it wrong, and ask for a redress. I would have done this long time ago.

"What misled you to carry your action against me was the same thing that misled me, leading to my problem. Both of us had limited insight, and it was this that led to our mistakes. My problem which everyone knows, how did it arise? Was it not because I lacked the insight to perceive how to handle properly what the creator of all things graciously gave me? In the middle of the problem, did not the creator step in and showed me his mercy? Did he not open my eyes to see what to do for a solution of the problem? The divine gift which people know me as having, is it not a gift from the creator, despite my mistakes? Who am I then, that I should turn my back on someone for his mistake, when the merciful creator didn't turn his back on me for my own mistakes? As he visited me and showed me his mercy, forgiving me my failings, so do I forgive you and welcome you with all my heart."

Immediately he finished his address, the men jubilated.

"You agree to the reconciliation," Killingworth said to Ward. "We request you to mark it with a visit to Oakland. We want all our people to witness this so that they will believe that you accepted our apology."

"When do you want me to visit?" asked Ward.

"We will fix the date and then inform you," replied Killingworth.

"I will visit any time you want me to visit," Ward promised. He then dashed out. Following this was Ward's household carrying food from the kitchen and beer from the nearby hut to the men. It was a great feast. The men ate and drank to their fill, thanked Ward and then went home happily.

After some time, Ackley and Brent brought a message to Ward. Oakland people had arranged a festival to take place at the forthcoming full moon of the summer season. Ward and his family were invited as special guests. When the day came, they attended and saw a big statue erected in the community centre. It was an effigy of Ward, a symbol of mystical insight, healing power and benevolence. For four days and four nights, under the summer sun and moon, Ward and his family were hosted by the people and entertained with food, drink and music. After this, the festival became an annual event, even to this day, taking place every summer in commemoration of Ward's achievement in Oakland.

By now, Oakland had joined Ward's friends in announcing far and wide his innocence and virtue. In addition to the spread of the story about his wealth and healing power, the news about his deep insight into the mysteries of nature had reached the uttermost parts of the land. In the whole territory, Ward was known as a mystic who was in regular communication with the gods. Even far away, some saw him as a deity.

Chapter 12

It was autumn season, the approach of nature's night. Rounding off its annual function, weary, the living earth was dozing, heading to slumber. In readiness for hibernation, hedgehogs fed, accumulating fat, and plants shed their golden leaves. While birds were migrating away, squirrels gathered nuts, storing them in their barns.

Ward's life was in tune with the rhythm. His crop had grown and yielded abundant light brown grain. He had done the autumn harvest and filled his barn with the grain. Already he had enough store of livestock feed, and logs of firewood for the winter ahead.

It was in this period that Ward wanted to start paying tribute to Wealdhere, the king of Owldown. Owldown was the political, commercial and spiritual centre of the whole territory. In it was a majestic Stonehenge which was part of arrays of its religious monuments. During the solstice, people from far and wide came to the Stonehenge to worship and watch in awe a mystifying phenomenon – the sun in a unique position where it seemed to act as a messenger of the sky god, sending its rays like a mystical information to the sacred Stonehenge.

Owldown priests, astrologers, and oracles were highly revered. People came to the astrologers to receive astrological information. They also came for the priests and oracles to help them offer sacrifices to the fertility goddess and sky god, and to communicate with the spirits of their ancestors, and the spirits of their relatives who had passed away. Owldown had powerful soldiers, and all the communities in the territory were its protectorates. At its gate was an awe-inspiring giant image of an owl which was a symbol of Owldown's mystical and protective power. Ward deeply respected Owldown for these.

But there was an event that enhanced Ward's adoration of Owldown. One day, he was in Owldown market to exchange his agricultural produce for more draft animals. As he was there, suddenly he heard an outburst of jubilation. The king of Owldown was returning from settling a conflict between Dunnockforest and Fiercegale tribes. His escort was Owldown soldiers on their horses. Having heard that the king had successfully resolved the conflict, the people in the market burst out in jubilation, thanking, waving at him.

Ward also heard from the people that many distant communities who were trading with Owldown were enemies of each other before, but later were reconciled through the King's mediation. After the reconciliation, the king facilitated a trade agreement between Owldown and the communities. Since then, Owldown and the communities were like a united society whose trade centre was Owldown market.

Ward learnt a lot from what he heard. He understood the reason behind the tributes which Owldown protectorates paid to the king. He knew that Floradale was part of the communities which Owldown protected. So, he wanted to contribute to strengthening the relationships between the communities and Owldown. He wanted to start this by paying tribute to the king. So, in that autumn season, he set apart some of his livestock and a good number of bags of grain. One night, he called Willow and Edmund.

"I called you to make you aware that one's good neighbour is his friend, and a friend is one of the sources of one's well-being," he said. "This is demonstrated by the relationship between us and Oakland. You don't have to be told how painful our life was when the relationship was broken, and how enjoyable our life is after the restoration of the relationship. Recently, I became aware of the importance of Owldown in the life and well-being of communities in the whole territory and even tribes far away. Owldown protects these communities and tribes, thereby ensuring their existence. It facilitates their good relationship, thereby ensuring their peace and well-being. These communities and tribes pay tribute to the king of Owldown in appreciation of the role the king plays. As we are part of the communities, it is therefore wise for me to join them in paying the tribute."

Willow and Edmund were very happy about this. They thanked Ward so much and agreed with him.

"When are you going to start this?" Willow asked.

"I will start it tomorrow," Ward replied. "I will like to go with Edmund so that after giving the king the tribute, I will introduce him to the king."

Edmund was very happy because it was an opportunity for him to see the splendour of the king's palace. So, he enthusiastically agreed to go with his father.

After the discussion, they went to bed, enjoying their sweet sleep. But while they were thus enjoying, a wind of disaster was blowing and the earth trembled in Owldown. The radiant sun was losing its light, sinking down the horizon, rendering the ancient world a gloomy void. Consternation had engulfed the people of Owldown as the pillar of their society was crumbling, collapsing. The mighty had fallen. Struck by sudden illness, Owldown's king was at the brink of untimely joining his ancestors.

That night Wealdhere collapsed, the night darkness vanished in the presence of the torchlight of Owldown herbalists, oracles and priests who were hastening in shock to the king's place to carry out herbal treatment and ritual healing. When they arrived, they saw Owldown noble men waiting for them. One of them was Egbert. The healers had tried their best but failed to heal the king.

"Sons of Owldown," Egbert said in sorrowful mood, and with unsteady voice, addressing the men, "although I am not a healer, I am one of the oldest men in Owldown. With my experience as an old man, I can tell you that following the way I am seeing things, Wealdhere is about to depart this life." he paused with a feeling of utter hopelessness, and the crowed shuddered in apprehension.

"The sign of illness he is suffering from," he continued, "is it not the same sign of the illness that killed his grandfather? Did not our healers try their best to heal him? Did they succeed? No, In the end the grandfather died.

"Our healers have done their best and yet the king's health continues to worsen. If they can't save the situation, who else can? In the whole territory, who is the healer that is better than ours? There is none. I must tell you that Wealdhere is about to depart this life. There is nothing else we can do now other than start planning for a burial," he paused helplessly. The crowd gaped at him in despair.

"I know a man who can heal him," a sharp voice pierced the air. The men immediately turned their attention to the speaker. It was a youth, Keenstar standing at a corner.

"Who is the man?" Willoughby asked sharply.

"He is called Ward," replied Keenstar.

"Where does he live?" Kennard asked calmly.

"It is very far from here," the youth replied.

"Where is it?" Egbert asked gently.

"It is near Woodland," Keenstar replied.

There was an outburst of scolding and mockery from some of the men. They presumed that the man Keenstar meant was Wood, the oracle of Woodland.

"Is he not the son of Spears?" Willoughby shouted angrily. He got up in anger and continued. "A true image of his father. This is how his father behaved. All the communities whose tributes and respect come to us, the father always went to them and disgraced us, telling them that we are not what they think we are. Always he went to destitute communities, working as a slave for them, returning with chickenfeed in exchange. This wandering is what this boy inherited from his father. He has wandered into the hinterland and can't keep it to himself, but dares to come here and announce it." Turning directly to the boy, he continued, "Yes lad, go on. Tell us to join you in wandering like a fugitive. Tell us to go to poverty-stricken hinterland inhabitants and beg them for food. Let's go to them and tell them that we don't have farmers to feed us. Let's go and beg them to feed us with their wild grain. Go ahead, go and tell Wood that he is better than our healers, and that we need him to heal our king".

As he spoke, some of the men mumbled to each other in mixed bitterness and disappointment. "But who brought him here?" They asked.

Keenstar was so frightened that he was about to sneak out.

"Stay, don't go!" Egbert ordered the boy before speaking to the men. "Which of you here in his heart of hearts will not listen to anything that might lead to a prevention of this imminent disaster?" he asked in sincerity. "Which of you here is so insensitive that he will not be ready to beg the poor for anything that will ensure the existence of our society? Who among us here will not be ready to bear people's mockery and laughter in search of a means to save our king?" Suddenly, he turned to Keenstar.

"My son," he said calmly and slowly, "you saw Wood of Woodland healing someone? Tell us the one he healed and the type of illness the one had."

"The man I mean is not Wood, neither is he from Woodland," replied Keenstar. "His name is Ward. He lives in a place called Floradale. The place is said to be somewhere around the hinterland."

"Who was the one he healed," Egbert asked cautiously.

"He healed many people," replied Keenstar. "One of them was a man from Mooreland. He was suffering from an illness similar to that of the king. Everyone thought that he was going to die. But when Ward was called, he healed him."

The crowd looked at each other in surprise.

"Do you know a way to Floradale?" Egbert asked.

"I haven't been there," replied Keenstar, "but I saw the man several times in Oakland. I know his father-in-law."

"Who is he?" asked Egbert.

"He is Woodrow of Oakland," replied Keenstar.

The men murmured to each other in surprise, as they remembered the prominent man.

Egbert breathed a sigh of relief and turned to the crowd and pleadingly spoke with a loud voice. "Elders, sons of Owldown, let's not miss this chance. Whatever it costs, we can afford it."

On finishing the statement, some of the men got up and dashed out. Very soon they returned with some soldiers and horses. Egbert addressed the soldiers:

> "Owldown and the whole territory are in trouble and so need your help," he said in persuasion. "This is the time your speed in horse riding is most needed. Go to a healer called Ward who this child told us about." He pointed at Keenstar as he spoke. "Tell him that Owldown needs his help. We desperately need him to heal our king. Tell him the health condition of Wealdhere. Tell him that whatever he asks for, in exchange for his healing the king, that we will give him. Take Keenstar with you to Oakland where he will show you Woodrow who will take you to the healer." Turning to one of the soldiers, he instructed: "Marsh, I trust you as a good leader. Take the lead and hurry to Oakland."

Mounting the horses with Keenstar, the soldiers sped to Oakland. They arrived in the morning, heading towards Woodrow's homestead. Woodrow heard the sound of the galloping horses, so, he walked towards the men. He noticed that the soldiers on the horses were natives of Owldown. He was puzzled as to why there was anxiety on their faces.

"What is wrong?" he asked the men, in surprise.

"Our king is very ill and we need Ward urgently for his medicine," Marsh replied hurriedly.

Quickly, Woodrow ran to one of the huts in the homestead and soon returned with Brent.

"What is happening to the King?" Brent asked in shudder.

"He is seriously ill," replied Marsh. "We come for Ward's help."

In shock, Brent ran very fast to his horse. Mounting it, he and the rest sped to Floradale. On their way, they saw Ward and Edmund who were taking the tribute in their cart to Owldown.

"Ward help!" Brent shouted as they approached.

"What is the matter?" Ward asked in surprise.

"Owldown's king is seriously ill," Brent replied as he panted.

"What illness is that." Ward asked in shock.

"Ask the soldiers, they have the details," replied Brent.

"Is the King seriously ill as Brent said?" Ward asked the soldiers in disbelief. "What type of illness is that and how did it start?"

"At the first cockcrow in the night, he collapsed, following a sudden unknown illness," Marsh replied in despair. "The illness seemed to affect his chest. He has breathing difficulty. Our healers have tried to heal him but failed. We are sent to seek your help."

On hearing this, Ward was terrified. It was a serious illness indeed. He quickly turned back, and all of them galloped to Floradale. On their arrival, he took his machete, hoe, sickle and a bag and rushed to the forest. Soon, he returned with a bagful of leaves, stems, roots and wood bark. Some of them were fresh and some dry. With these, he and the men on their horses sped to Owldown.

Having arrived at the king's palace, Ward went to his bedside and examined him. He then took the medicinal plants to a room which was made available for him. He extracted the sap of some of the plants, and ground some of the bark and the dry ingredients into powder. He mixed these, forming concoction. He took the concoction to the patient and started the healing. First, he recited an incantation designed to make the patient's body receptive to the treatment. He then gave some of the medicine to the patient through his mouth and rubbed some on him.

As time passed, the treatment began to yield positive results. The patient started breathing normally and his body moved freely. The clutches of death were broken and he was set free. He became conscious of his surroundings and he communicated with people around him. Noise of relief, joy and excitement from the people filled the air.

"This is not yet a full cure," Ward said to the men in the middle of their excitement. "It is only a resuscitation." He then went back to the medicine room and later returned with a different concoction and continued the treatment.

For days, Ward stayed in Owldown, healing and nursing the king. Later, he went to Floradale and came back with more medicinal plants. That was for speeding up the patient's full recovery. He handed them to Owldown herbalists. After instructing them how to administer the medicine, he returned to Floradale.

Shortly, Ward received a message from the people of Owldown that he and his family were invited to join them in celebrating the healing of the king. When the day came, they attended and saw Owldown alive with music and dance. They were warmly received and taken to the king's palace for a banquet where they ate on the same table with the king and the noble men of Owldown. After the banquet, the king addressed Ward in the presence of the men.

"Ward," the King called in a happy mood, "we are happy to see you and your family here. It gladdens our hearts that what we lacked in our land, we have it today and that is you. Through healing me, you proved this. You proved that you deserve our tribute. People in the whole territory bring tribute to me in appreciation of what they feel I do for them. If I accept the tribute from them, I will feel guilty for not paying you a tribute, because what you did is far greater than what I do for the people."

Then the king and his men took Ward outside and showed him many bags of grain, new agricultural tools, many livestock, including draft animals.

"These are our gift to you for what you did," the king said to Ward. "In addition, from today on, you and your family are members of our community. We have recognised Floradale and the clans and hamlets around it as a unified community. We have a regular meeting in Owldown where emissaries from different communities come on behalf of their communities for putting ideas together, for preserving peace and order in our society. You will be the emissary who will be coming on behalf of your community, just as Woodrow is the emissary that comes on behalf of Oakland.

"We heard of your deep insight into the mysteries of nature. In our annual remembrance day, people with outstanding knowledge and experience give a talk about their knowledge and experiences, and people from different places come to hear them for improving their own knowledge. Considering your experience and insight, we agreed that you will be one of those who will be giving a talk in the rememberance day."

Ward was overjoyed. He bowed down to the king and thanked him and his men.

Then the men called some of their soldiers and farmers to take the gift, Ward and his family to Floradale.

Soon, the remembrance day came. There were different events that took place on the occasion. The major one was a ceremonial activity in celebration of their victory in the Great War they fought in the past. It was a war in which a myriad of flying arrows and spears almost eclipsed the sun. The other was the talk presentation.

Ward had arrived at the king's palace to give the talk, and wise men from different places were there as his audience. He gave a detailed talk about his experience as a farmer and how he gained insight into the interrelationships of organisms and the significance. He elaborated on how interactions of organisms with each other and their surroundings help to maintain and perpetuate the conditions for life on the earth, and how the diversity and unique characteristics of organisms make this possible. He informed that man evolved as a custodian whose role is to take care of the organisms, and in doing so, he achieves the purpose of his existence on the earth. He closed the talk with thanks to the audience for their attention.

Astonished, the audience applauded. The wise men from Brightfort tribe who made a long journey to the event were amazed and extremely happy. Their leader, Way got up and spoke in overwhelming feeling of joy:

> "Oh Owldown!" his voice rose in excitement and praise. "My eyes have seen. My ears have heard. Today we have got what we have been endlessly searching for, for so long.

"Owldown! We have witnessed that you are the land which our prophets saw as the place where the information which our goddess gave us will be confirmed as valid information. We can now go home and rest happily, announce to our people that the prophecy is fulfilled; that we got what we have been travelling, searching for; that in Owldown, our goddess was proved to be the most truthful of all deities."

Way's words were not just a praise of Owldown, but also an announcement that Owldown was a land chosen by the gods for divine revelation. For this, Ward was held in high reverence by the people of Owldown.

It was many years ago when this happened. By now, Ward was very old and couldn't anymore carry out many of his usual duties. The duties had been taken over by Edmund, including the roles he used to play in Owldown.

Ward lived to see the birth and growth of his great grandchildren. One day, he called members of his family and spoke to them.

"Edmund," he called. "I am happy that you are doing well in carrying out all the duties you took over from me. I beg all of you to support him in his effort. Also, I beg you to always obey Willow in whatever she tells you to do. It is by heeding her advice that I was able to achieve many of the things I achieved, which you all enjoy." He paused and suddenly raised his face, looking at the horizon. There was sadness in his mood.

"Where are they heading to, and what does the future hold?" he cried out. "Invasion and counter-invasion everywhere. War rising beyond imagination. Men creating hunger, warfare and suffering by beating their ploughshares into deadly weapons for destroying life on the earth."

Stunned, the family gaped at him, presuming that he was hallucinating. But the future would tell if he was hallucinating or otherwise prophesying.

Then Ward turned to the family and instructed: "Do your best to support Owldown in its work for all the communities. This includes its facilitation of the communities' coexistence. This is a divine duty required of man."

Shortly after the address, Ward joined his ancestors. Ward had accomplished his mission, demonstrating man's origin and central role on earth – as one who rose from the ashes and equipped to manage nature, care for life on the earth, and through this, he immortalised his name and soul, and left behind an enlightening legacy. The legacy which Ward left behind lasted very long, and part of it is the knowledge of the interrelationships of organisms and the significance – an idea which was the forerunner of the theory and principle of ecosystem.

www.ingramcontent.com/pod-product-compliance
Ingram Content Group UK Ltd.
Pitfield, Milton Keynes, MK11 3LW, UK
UKHW042059020225
454580UK00001B/15